AF874099

Studies in Systems, Decision and Control

Volume 197

Series Editor

Janusz Kacprzyk, Systems Research Institute, Polish Academy of Sciences, Warsaw, Poland

The series "Studies in Systems, Decision and Control" (SSDC) covers both new developments and advances, as well as the state of the art, in the various areas of broadly perceived systems, decision making and control–quickly, up to date and with a high quality. The intent is to cover the theory, applications, and perspectives on the state of the art and future developments relevant to systems, decision making, control, complex processes and related areas, as embedded in the fields of engineering, computer science, physics, economics, social and life sciences, as well as the paradigms and methodologies behind them. The series contains monographs, textbooks, lecture notes and edited volumes in systems, decision making and control spanning the areas of Cyber-Physical Systems, Autonomous Systems, Sensor Networks, Control Systems, Energy Systems, Automotive Systems, Biological Systems, Vehicular Networking and Connected Vehicles, Aerospace Systems, Automation, Manufacturing, Smart Grids, Nonlinear Systems, Power Systems, Robotics, Social Systems, Economic Systems and other. Of particular value to both the contributors and the readership are the short publication timeframe and the world-wide distribution and exposure which enable both a wide and rapid dissemination of research output.
** Indexing: The books of this series are submitted to ISI, SCOPUS, DBLP, Ulrichs, MathSciNet, Current Mathematical Publications, Mathematical Reviews, Zentralblatt Math: MetaPress and Springerlink.

More information about this series at http://www.springer.com/series/13304

Krzysztof Patan

Robust and Fault-Tolerant Control

Neural-Network-Based Solutions

Springer

Krzysztof Patan
Institute of Control and Computation
Engineering
University of Zielona Góra
Zielona Góra, Poland

ISSN 2198-4182 ISSN 2198-4190 (electronic)
Studies in Systems, Decision and Control
ISBN 978-3-030-11868-6 ISBN 978-3-030-11869-3 (eBook)
https://doi.org/10.1007/978-3-030-11869-3

Library of Congress Control Number: 2018967749

MATLAB® and Simulink® are registered trademarks of The MathWorks, Inc., 1 Apple Hill Drive, Natick, MA 01760-2098, USA, http://www.mathworks.com.

© Springer Nature Switzerland AG 2019
This work is subject to copyright. All rights are reserved by the Publisher, whether the whole or part of the material is concerned, specifically the rights of translation, reprinting, reuse of illustrations, recitation, broadcasting, reproduction on microfilms or in any other physical way, and transmission or information storage and retrieval, electronic adaptation, computer software, or by similar or dissimilar methodology now known or hereafter developed.
The use of general descriptive names, registered names, trademarks, service marks, etc. in this publication does not imply, even in the absence of a specific statement, that such names are exempt from the relevant protective laws and regulations and therefore free for general use.
The publisher, the authors, and the editors are safe to assume that the advice and information in this book are believed to be true and accurate at the date of publication. Neither the publisher nor the authors or the editors give a warranty, express or implied, with respect to the material contained herein or for any errors or omissions that may have been made. The publisher remains neutral with regard to jurisdictional claims in published maps and institutional affiliations.

This Springer imprint is published by the registered company Springer Nature Switzerland AG
The registered company address is: Gewerbestrasse 11, 6330 Cham, Switzerland

To my family

Foreword

This monograph aims at presenting novel ideas, concepts and results in robust fault-tolerant control. Rapid developments in control technology have an impact on all the areas of the control discipline: there emerge new theories, advanced control solutions, new industrial processes, computer methods and implementations, new applications, new philosophies, and, inescapably, new challenges. Much of this development work is presented in the form of industrial reports, feasibility study papers and reports on advanced collaborative projects. Therefore, this monograph offers an opportunity for researchers, practitioners and students to gain access to an extended and clear exposition of new investigations in all the aspects of robust fault-tolerant control, intended for a rapid dissemination of the results and accessible to a wider readership.

As many technological systems are becoming increasingly complex, more widespread and integrated, the effects of system faults can potentially be devastating to the infrastructure of any modem society. Feedback control is just one important component of the total system supervision. Fault-tolerant control describes another set of components having extensive commercial, industrial and societal implications; it is imperative, however, that we are able to make use of them in a robust and inexpensive manner. The model-based approach is the usual solution of the practical fault-tolerant control design but, as the author Krzysztof Patan has highlighted in this monograph, the methodologies based on neural networks can also be successfully exploited. The search for reliable, robust and inexpensive fault-tolerant control methods has been ongoing since the early 1980s. Since 1991, the SAFEPROCESS Steering Committee, created by the International Federation of Automatic Control (IFAC), is in operation promoting research, developments and applications in the field of fault-tolerant control. The last decade has seen the formalisation of several theoretical approaches accompanied by some attempts to standardise the nomenclature in the field.

There are not many research publications within this important research area: one can point to certain monographs that can be said to provide interesting contributions to fault-tolerant control describing, however, the topic from slightly different points of view. To these, we can now add this monograph by Krzysztof

Patan. The key features of this text include a useful survey material, a description of new approaches (utilising data-driven and neural-network-based methodologies), as well as a number of experimental studies helpful in understanding the advantages and the drawbacks of the suggested strategies and tools. Different groups of readers, from industrial engineers wishing to gain insight into the applications potential of new fault-tolerant control methods relying on artificial intelligence tools, to the academic control community looking for new problems to tackle will find much to learn from this monograph.

Ferrara, Italy
October 2018

Silvio Simani

Preface

Indisputably, what is known as the robust and the fault-tolerant approaches have become important and essential subclasses of modern control theory. Nowadays, control systems designed for industrial plants have to meet the high requirements for the operation safety, stability and control performance. The notion of system robustness is made more concrete by means of the following two important notions. *Robust stability* means that the system remains stable for every plant belonging to the uncertainty set, whereas *robust performance* means that the performance specifications are satisfied for every plant belonging to the uncertainty set. Arguably, both of these are some of the most desirable features of the designed control systems. Robustness, however, is a problem that is hard to solve in the context of nonlinear systems. While robust control strategies allow a system to cope with model uncertainty, *fault-tolerant* control allows the system to cope with possible faulty situations occurring in industrial plants. The main objective of fault-tolerant control is to continue the plant operation, possibly at a reduced performance, and to preserve stability conditions in the presence of unexpected changes of system work caused by faults. There are, however, many problems encountered when designing fault-tolerant control for nonlinear systems.

Solutions of both robust control and fault-tolerant control problems can be obtained through the use of artificial neural networks. Neural networks can be effectively applied to deal with uncertainty modelling for the robust control purposes as well as to design the fault diagnosis units required by fault-tolerant control. The book proposes a number of strategies based on neural networks for nonlinear systems, e.g. model predictive control, control reconfiguration approaches and iterative learning control. Each proposed control strategy is accompanied by an example showing its applicability.

The material included in the monograph results from research that has been carried out by the author at the Institute of Control and Computation Engineering (the University of Zielona Góra, Poland) for the last eight years in the area of the modelling of nonlinear dynamic processes as well as control of industrial processes. Some of the presented results were developed with the partial support of the Ministry of Science and Higher Education in Poland under the grants N N514

678440 *Predictive fault tolerant control for nonlinear systems* (2011–2014), 2014/15/B/ST7/03208 *Improvement of the control performance using iterative learning* (2015–2018) and 2017/27/B/ST7/01874 *Learning-based methods for high-performance robust control* (2018–2021).

The monograph is divided into seven chapters. Chapter 1 introduces the subject matter. Chapter 2 is a survey of artificial neural networks that have possible applications to modelling and control. Some space is also devoted to the important problems of model training and the development of robust models. Chapter 3 describes the notion of control systems synthesis, focusing on the role of neural networks in that context. We also highlight the notions of robust and fault-tolerant control. Chapter 4 presents the model of predictive control based on neural networks. Fault tolerance as well as robustness of the proposed nonlinear predictive schemes are also discussed there. Chapter 5 presents the fault accommodation and control reconfiguration approach where neural networks are used in the following ways: (1) to process modelling; (2) to design what is known as a nonlinear observer and (3) to aid in uncertainty modelling. Chapter 6 discusses a number of methods that make use of neural networks in the context of iterative learning control with an emphasis on the problems of convergence and stability. Finally, Chap. 7 presents our contribution to the area of control in the context of industrial processes.

At this point, I would like to express my sincere thanks to all the colleagues from the Institute of Control and Computation Engineering at the University of Zielona Góra for many stimulating discussions and a friendly atmosphere, which was a big factor in my success in writing up this monograph. In particular, I would like to thank my former Ph.D. student Andrzej Czajkowski for his contribution to Chap. 5, my brother Maciek for his contribution to Chap. 6, and Wojtek Paszke who pointed my attention to the area of iterative learning control. Finally, I would like to express my gratitude to Dr. Adam Trybus for proofreading the text and providing linguistic advice.

Zielona Góra, Poland
September 2018

Krzysztof Patan

Acknowledgements

The ideas on robust and fault-tolerant control presented in this monograph were developed over the last few years and have previously appeared in a number of publications. However, the purpose of the book is to provide a unified presentation of these solutions and bring them together in a single publication. In order to achieve this objective, it has been necessary at times to reuse some material that we published in earlier works. In spite of the fact that such material has been modified, expanded and rewritten for the monograph, permission from the following publishers is acknowledged.

Springer, Berlin is acknowledged for permission to reuse portions of the following chapters.

Krzysztof Patan, Locally Recurrent Neural Networks of *Artificial Neural Networks for the Modelling and Fault Diagnosis of Technical Processes*, vol. 377 in the series Lecture Notes in Control and Information Sciences, Springer-Verlag, Berlin, 2008.

Acknowledgement is given to the Institute of Electrical and Electronic Engineers for permission to reproduce parts of the following papers.

Krzysztof Patan, Neural Network-Based Model Predictive Control: Fault Tolerance and Stability, *IEEE Transactions on Control Systems Technology*, vol. 23, no. 3, pp. 1147–1155, 2015.

Krzysztof Patan, Maciej Patan, Damian Kowalów, Optimum training design for neural network in synthesis of robust model predictive control, in *Proceedings of 55th IEEE Conference on Decision and Control*, Las Vegas, USA, pp. 3401–3406, 2016.

We acknowledge the permission of the Elsevier to reproduce portions of the following papers.

Andrzej Czajkowski, Krzysztof Patan, Mirosław Szymański, Application of the state-space neural network to the fault-tolerant control system of the PLC-controlled laboratory stand, *Engineering Applications of Artificial Intelligence*, vol. 30, pp. 168–178, 2014.

Krzysztof Patan, Two stage neural network modelling for robust model predictive control, *ISA Transactions*, vol. 72, pp. 56–65, 2018.

Zielona Góra, Poland
October 2018

Krzysztof Patan

Contents

Notation

Symbols

$\mathbb{R}$	Set of real numbers
$\mathbb{N}$	Set of nonnegative integers
t, k	Continuous and discrete-time indexes
p	Trial index
$y(\cdot)$, $\hat{y}(\cdot)$	System output and estimated system output
$u(\cdot)$, $\hat{u}(\cdot)$	System input and estimated system input
$x(\cdot)$, $\hat{x}(\cdot)$, $\bar{x}(\cdot)$	State vector, estimated state and nominal state
$u(\cdot)$	Input vector
$y(\cdot)$	Output vector
$\sigma(\cdot)$, $\boldsymbol{\sigma}(\cdot)$	Activation function and vector-valued activation function
J	Cost function
$\boldsymbol{A}$	State matrix
$\boldsymbol{W}$	Weight matrix
$\boldsymbol{C}$	Output matrix
$\boldsymbol{B}$	Input (control) matrix
$\boldsymbol{D}$	Transfer matrix
$\mathcal{N}(m, \sigma)$	Normally distributed random number with the expectation value m and the standard deviation σ
α	Significance level
t_α	Tabulated value assigned to the significance level α
C^1	Class of continuously differentiable mappings
N_p, N_u, N_c, N_y	Prediction horizon, control horizon, constraint horizon and output constraint horizon
T_s	Sampling time
$\boldsymbol{I}$	Identity matrix
$\boldsymbol{0}$	Zero matrix

$\varphi(\cdot)$, $\phi(\cdot)$	Regression vectors
L	Lipschitz constant
$e(\cdot)$	Tracking error

Operators

∇	Gradient
∂	Partial derivative
T	Matrix transposition
sup	Least upper bound (supremum)
inf	Greatest lower bound (infimum)
max	Maximum
min	Minimum
arg max	Argument of a maximum value
arg min	Argument of a minimum value
rank($\boldsymbol{A}$)	Rank of a matrix $\boldsymbol{A}$
det($\boldsymbol{A}$)	Determinant of a matrix $\boldsymbol{A}$
trace($\boldsymbol{A}$)	Trace of a matrix $\boldsymbol{A}$
$\|\boldsymbol{w}\|$	Vector norm
$\|W\|$	Matrix norm
W^{-}	Matrix pseudoinverse

Abbreviations

ANN	Artificial Neural Network
ARX	Auto-Regressive with eXogenous input
BDM	Binary Diagnostic Matrix
BP	Back-Propagation
BPTT	Back-Propagation Through Time
CMPC	Constrained Model Predictive Control
CRHPC	Constrained Receding Horizon Predictive Control
DBN	Deep Belief Network
DFT	Discrete Fourier Transform
ESN	Echo State Network
FD	Fault Diagnosis
FDI	Fault Detection and Isolation
FIM	Fisher Information Matrix
FIR	Finite Impulse Response
FPE	Final Prediction Error
FSS	Feasible System Set
FTC	Fault-Tolerant Control

GMDH	Group Method and Data Handling
GPC	Generalised Predictive Control
IF	Integrate and Fire
IIR	Infinite Impulse Response
ILC	Iterative Learning Control
IMC	Internal Model Control
KKT	Karush–Kuhn–Tucker
LM	Levenberg–Marquardt
LMI	Linear Matrix Inequality
LQ	Linear Quadratic
LRGF	Locally Recurrent Globally Feed-forward
LSTM	Long Short-Term Memory
LVQ	Learning Vector Quantization
MDM	Multivalued Diagnostic Matrix
MEM	Model Error Modelling
MPC	Model Predictive Control
MPCD	Model Predictive Control with Disturbance model
MRAC	Model Reference Adaptive Control
MSE	Mean Square Error
NAR	Nonlinear Auto-Regressive
NARMAX	Nonlinear Auto-Regressive Moving Average with eXogenous input
NARX	Nonlinear Auto-Regressive with eXogenous input
NFIR	Nonlinear Finite Impulse Response
NIIR	Nonlinear Infinite Impulse Response
NLARX	NonLinear wavelet Autoregressive Regressive with eXogernous input
NOE	Nonlinear Output Error
OED	Optimum Experimental Design
PCNN	Pulse-Coupled Neural Network
PD	Proportional Derivative
PI	Proportional Integral
PID	Proportional Integral Derivative
PNN	Probabilistic Neural Network
RBF	Radial Basis Function
RBM	Restricted Boltzmann Machine
RMLP	Recurrent Multi-Layer Perceptron
RMPC	Robust Model Predictive Control
RNN	Recurrent Neural Network
RTRL	Real-Time Recurrent Learning
RTRN	Real-Time Recurrent Network
SGPC	Stable Generalised Predictive Control
SM	Set Membership
SNN	Spiking Neural Networks
SOM	Self-Organising Map
SSE	Sum of Squared Errors

SSIF	State-Space Innovation Form
SSNN	State-Space Neural Network
TDL	Tapped Delay Line
TDNN	Time Delay Neural Network
TDRBP	Time-Dependent Recurrent Back-Propagation

List of Figures

List of Tables

Chapter 1
Introduction

1.1 Scope of the Book

Industrial automation is a field where significant developments occur at an ever-increasing pace. Nowadays, it is not only very sophisticated and complex industrial plants that are fully automated but also an increasing number of simpler designs fall into that category as well. In order to satisfy the high demands for control quality, a number of sophisticated control strategies have been proposed. This includes what is known as *model-predictive* control (MPC), *adaptive* control, *optimal* control and *iterative-learning* control (ILC). Most of the existing control strategies are model based. Therefore, a crucial stage of the control system synthesis relates to a proper identification of the plant to be controlled. For linear systems both identification and control system design have reached a high level of maturity. However, there is still a need for developing identification methods and control algorithms for nonlinear systems. To date, many methods for nonlinear system identification have been proposed, e.g. Volterra series, the group method and the data handling (GMDH) models, Wiener and Hammerstein models, nonlinear auto-regressive models or dynamic neural networks [11, 22]. Undoubtedly, neural networks are the most popular models to be used in that context. These are very flexible, universal and are often used in cases when there is no accurate mathematical model of the process at hand. The first dynamic neural network models were developed in the early 1980s and the recurrent networks proved their usefulness in the control area in both modelling [6, 19, 21] and control [12, 20] of nonlinear dynamic processes. Artificial neural networks provide an excellent mathematical tool for dealing with nonlinear problems. They have an important property, namely that any continuous nonlinear relationship can be approximated with arbitrary accuracy using a neural network with a suitable architecture and weight parameters [5, 6]. Furthermore, a neural network can extract the system features from the historical training data using a learning algorithm, requiring little or no a priori knowledge about the system. This means that the modelling of nonlinear systems can be more flexible, especially in the cases where there is no possibility

© Springer Nature Switzerland AG 2019

K. Patan, *Robust and Fault-Tolerant Control*, Studies in Systems, Decision and Control 197, https://doi.org/10.1007/978-3-030-11869-3_1

of figuring out the analytical input-output representation of the system. All these characteristics make neural networks well suited to control system synthesis.

In the field of modern control theory one can distinguish two intensively developed research avenues: robust control and fault-tolerant control (FTC). Real-life industrial systems abound in disturbances, distortions and noise affecting their performance. More importantly, such problems cannot be simply solved, neither can they, obviously, be neglected. We emphasise also that a model of the process used for control synthesis is not a perfect replica of the process dynamics. It follows that it is necessary to apply a control scheme taking the above factors into account and trying to minimize their impact on control quality. Thus, robust control is of crucial importance in feedback control [1, 24]. On the other hand, along with an increasing number of plant components, the probability that a fault occurs in the system is also increasing. This makes it necessary to implement a control system that automatically adapts and changes the behaviour of the entire system in case of unwanted abnormal situations. Such systems are referred to as fault-tolerant control. Sensor or actuator faults, product changes, material consumption can all affect the controller performance [11, 27, 28] and result in large financial losses or even in breaching the safety regulations [3, 11]. It should be stressed, however, that designing robust and fault-tolerant control systems for nonlinear plants is not a trivial problem. The existing solutions developed for linear systems cannot be applied directly. In many cases, the only possibility is to search for new approaches, including those, where the existing methods are being adapted to the new situation.

This monograph presents solutions for robust and fault-tolerant control that have general applicability to the entire class of nonlinear systems. Taking into account that control systems, fault-diagnosis (FD) algorithms as well as uncertainty estimates are mainly model based, and that our setting is nonlinear systems, we focus exclusively on the solutions employing neural networks.

Model-predictive control is a popular and frequently used control scheme, which showed its applicability in industrial contexts [7, 9, 13, 18, 25]. MPC has been the subject of intensive research for the last three decades [2, 14, 26]. The attractiveness of predictive control algorithms comes from their capability to take into account the process and technological constraints imposed on input, output or state variables. The second very important reason is that their operating principles are understandable and relatively easy to explain to practitioners, which seems to be a crucial aspect during an implementation of a new control scheme in industry. An important part of MPC strategies is a model of the controlled process used to predict future plant states/outputs based on the past and current states/outputs as well as future control signals. These are calculated through minimization of the cost function, taking into account constraints. In many cases, constraints are of the form of inequalities featuring the process variables, e.g. input saturations. Not taking such nonlinearities into account can result in degraded performance of closed-loop control and might lead to stability problems. Therefore, the stability of the control system is a fundamental requirement. As pointed out in outstanding survey papers of Morari and Lee [17] as well as Mayne et al. [15], the predictive-control theory has reached high level of maturity, especially for linear systems. However, there is still a number of difficulties when dealing with the nonlinear case. Problems are still observed with the modelling of nonlinear processes, state estimation, fault diagnosis or fault-tolerant control [17]. Also, the robustness of MPC against model uncertainty and noise is still a challenge. The robustness of a control system is referenced to a specific uncertainty range and specific stability and performance criteria. In spite of a rich literature devoted to a robust control of linear systems, further research is required to develop implementable robust controllers, especially for nonlinear plants [15]. The robustness of MPC can be achieved using a neural network to estimate uncertainty associated with the plant model. A data-driven method of deriving uncertainty can be seen as an interesting alternative to classical methods, such as structural or parametric uncertainty, which require uncertainty to be formulated in an analytical way. Data needed for neural network training can be easily acquired in the control system. From that point of view, solutions using neural networks can be seen superior to the structural or parametric ones, especially when nonlinear processes are considered. To date, the research on incorporating neural networks for the task of robust model development in the context of MPC is rather scarce. Our book is an attempt to fill this gap.

MPC seems to be also suitable for FTC, since representing both fault and control objectives in MPC is relatively simple. Certain faults can be represented through the modification of constraints in the MPC algorithm while others can be handled through modification of a system model [2, 10]. In terms of actuator or process faults, the predictive control should react in the standard way, whereas sensor faults should be treated differently. For example, if we consider a fault occurring in the output sensor, the control system immediately reacts on the change in the system output, which is the direct consequence of measuring a value that is outside the accepted range. However, in that case, a control signal value should be kept the same as before the fault occurrence, because such a fault does not change the output signal at all. Thus, the estimation of the size of a sensor fault is very important during the design phase of fault-tolerant predictive control.

In many cases, constraints are of the form of inequalities imposed on the process variables, e.g. the input saturations. If such nonlinearities are not taken into account, it may result in a degraded performance of closed-loop control and can lead to stability problems. Therefore, the stability of the control system is a fundamental requirement. Although the idea of MPC using neural networks is not new [20], the research connected with the stability of MPC based on input-output neural network models is rather scarce. Given the above, it seems desirable that the problem of stability of the predictive scheme should be further investigated. One of the possible solutions is to investigate whether a cost function is monotonically decreasing with respect to time, analogously as shown in [23]. In such a context, the derived stability conditions can be employed to redefine a constrained optimization problem in order to calculate control.

In spite of the fact that many effective control schemes have been described in the literature, including optimal, predictive or adaptive control, many industrial installations and plants still use the standard proportional-integral-derivative (PID) controllers. In such cases, fault-tolerant control system should be designed based on the existing control scheme with a feedback controller, e.g. of the PID type. In many situations, feedback control can prevent faults from being observed. Therefore, it is advisable to equip modern control systems with self-diagnostic mechanisms in order to manage abnormal situations in an automatic fashion. Fortunately, there is an exhaustive literature on model-based fault diagnosis [3, 8, 11]. The idea of fault compensation is quite simple. First, a fault estimate should be derived. This can be done based on the model of the plant and the available measurements. If a system variable is not available, a nonlinear observer should also be designed. Next, it is necessary to investigate the impact of the fault on the control signal. The problem of choosing an additional control signal to compensate the effect caused by a fault can be easily solved for linear systems. In such cases, the relation between the control and the fault is identified through some kind of an inverse model. Unfortunately, such approaches are much more difficult to develop for nonlinear processes. Moreover, in real-life plants due to noise or disturbances, the problem of determining an accurate model of a plant is also a challenge. Finally, as the process works in real-time and there are fixed time constraints imposed on the control algorithm, the computation burden is also a limitation to be taken into account. Designed in this way, fault-tolerant control starts to compensate the fault effect by adding an auxiliary signal to the standard control (derived by the classical feedback controller, e.g. of the PID type). This auxiliary control constitutes an additional control loop that can influence the stability of the entire system. Therefore, the stability of the proposed control scheme should be further investigated.

Originating in the late 1970s, iterative-learning control (ILC) has become a very popular control strategy for repetitive processes. ILC is the solution to the problem of perfect tracking in finite time under a repeatable control environment. The existing control algorithms are not able to satisfy such a requirement since they exhibit the same behaviour for each repetition (known also as a trial). The main idea behind ILC is to use a memory to store data from the current trial and use them in subsequent repetitions. As a result, the control system can learn from the previous trials how to improve the tracking performance in the future. For the class of linear time invariant system, the key stages of the analysis of ILC systems can be considered completed. For nonlinear systems there is still no unifying theory of the analysis and design of ILC system and such a result would be highly desirable [16]. In the literature, one can find certain interesting solutions applicable in the case of nonlinear systems, including affine and non-affine systems or in the case of linear-parameter-varying systems [4, 29]. However, in most of these, the authors used a linear iterative controller. As pointed out by Kevin Moore in his treatment of the topic [16], it makes sense to consider learning controllers that also have a nonlinear or time-varying structure. As artificial neural networks are able to represent a wide class of nonlinear systems, they can be considered a good candidate in terms of designing nonlinear ILC schemes. Despite the fact that the first solutions using neural networks in the context of ILC were proposed in the early 1990s, one can find only a handful of ILC systems realised by means on neural networks. This is mainly a result of problems with satisfying both convergence and stability conditions when a nonlinear iterative controller is used.

1.2 The Structure of the Book

The remainder of the book is organised as follows:

Neural networks. This chapter is devoted to the presentation of neural-network models in the context of control systems design. It is divided into four parts. The first two parts introduce the reader to the theory of static and dynamic neural network structures. These parts can be treated as a quick review of already developed and well-documented neural network architectures, giving an insight into their properties and the possibility of their application in control theory. The third part is focused on the problem of model design. As the majority of control system designs are model based, developing an accurate model of a plant is of crucial importance, especially for nonlinear systems. Two modelling approaches are discussed: forward and inverse modelling. Moreover, the problem of a training of feed-forward and recurrent neural models is described in the context of parallel and series-parallel identification schemes. The fourth part discusses a very important issue of uncertainty associated with the model. This notion is crucial when dealing with robust and fault-tolerant control. We describe the methods that could be used in estimating the uncertainty associated with neural network models, namely the set-membership identification, model error modelling and statistical approaches.

Robust and fault-tolerant control. The chapter introduces the reader into the theory of automatic control, focusing on nonlinear control schemes developed by means of artificial neural networks. It contains essential information on direct control based on neural networks, model reference adaptive control, feed-forward control, model-predictive control and optimal control. The role played by neural networks in each control scheme is emphasized. Since a desirable feature of modern control systems is some level of robustness and fault tolerance, the next two parts of the chapter discuss the problem of robust control as well as fault-tolerant control. The main objective of these parts is to present the existing approaches to achieving robustness and fault tolerance of control systems and to point out their drawbacks. It can be argued that the use of neural networks can help in solving a number of problems observed in the described methods. Some such solutions are presented and discussed in the following chapters.

Model-predictive control. The chapter contains the results of the original research dealing with robust and fault-tolerant predictive control schemes. The first part of the chapter is devoted to nonlinear predictive control developed by means of neural networks. Some of the most important issues connected with optimization and stability are investigated in detail. The next part introduces the sensor fault-tolerant control (For this purpose, predictive control is equipped with a fault-diagnosis block.) Binary diagnostic matrix as well as multivalued diagnostic matrix are used in this context. The proposed control strategy is tested using the tank unit example provided. We develop a robust version of predictive control based on a robust model of a plant. We investigate two approaches: uncertainty modelling using model error modelling (MEM) and statistical uncertainty estimation via statistical analysis. The proposed control schemes are tested on the example of a pneumatic servomechanism.

Control reconfiguration. The chapter is a description of our contribution in the form of algorithms for fault accommodation and control reconfiguration. The proposed FTC system detects a fault, estimates it and corrects a control law in order to compensate the fault effect observed in the control system. In order to create a control system, it is necessary to take into account the model of a plant as well as the state observer. Both models are designed using neural networks. As a result, the corrected control is obtained by an additional control loop that can influence the stability of the control system. Finally, the chapter also discusses the stability of the proposed control system. The proposed solutions are tested on the examples of a tank unit and two tank laboratory stands.

Iterative-learning control. The chapter presents original research results in the area of nonlinear iterative-learning control. We propose a novel ILC scheme developed using neural networks. The following two cases are described: dynamic and static learning controllers and in both cases the controller is designed in such a way as to minimize the tracking error. This task is accomplished by an appropriate training of the neural controller after each repetition of the control system. Additionally, the chapter contains both the stability and convergence analysis of the proposed nonlinear ILC. The portrayed control strategies are tested on the examples of a pneumatic servomechanism and a magnetic suspension system.

References

1. Åström, K.J., Kumar, P.R.: Control: a perspective. Automatica **50**, 3–43 (2014)
2. Camacho, E.F., Bordóns, C.: Model Predictive Control, 2nd edn. Springer, London (2007)
3. Chen, J., Patton, R.J.: Robust Model-Based Fault Diagnosis for Dynamic Systems. Kluwer, Berlin (1999)
4. Chen, Y., Wen, C.: Iterative Learning Control. Convergence, Robustness, Applications. Lecture Notes in Control and Information Sciences, vol. 248. Springer, London (1999)
5. Gupta, M.M., Jin, L., Homma, N.: Static and Dynamic Neural Networks. From Fundamentals to Advanced Theory. Wiley, New Jersey (2003)
6. Haykin, S.: Neural Networks. A Comprehensive Foundation, 2nd edn. Prentice-Hall, New Jersey (1999)
7. He, N., Shi, D., Forbes, M., Backstörm, J., Chen, T.: Robust tuning for machine-directional predictive control of MIMO paper-making processes. Control Eng. Pract. **55**, 1–12 (2016)
8. Isermann, R.: Fault Diagnosis Systems. An Introduction from Fault Detection to Fault Tolerance. Springer, New York (2006)
9. Janakiraman, V., Nguyen, X., Assanis, D.: An ELM based predictive control method for HCCI engines. Eng. Appl. Artif. Intell. **48**, 106–118 (2016)
10. Joosten, D.A., Maciejowski, J.: MPC design for fault-tolerant flight control purposes based upon an existing output feedback controller. In: Proceedings of 7th International Symposium on Fault Detection, Supervision and Safety of Technical Processes, SAFEPROCESS 2009 Barcelona, Spain, 30th June–3rd July 2009. CD-ROM
11. Korbicz, J., Kościelny, J., Kowalczuk, Z., Cholewa, W. (eds.): Fault Diagnosis. Models, Artificial Intelligence, Applications. Springer, Berlin (2004)

12. Ławryńczuk, M.: Computationally Efficient Model Predictive Control Algorithms. A Neural Network Approach. Studies in Systems, Decision and Control, vol. 3. Springer, Switzerland (2014)
13. Li, S., De Schutter, B., Wang, L., Gao, Z.: Robust model predictive control for train regulation in underground railway transportation. IEEE Trans. Control Syst. Technol. **24**, 1075–1083 (2016)
14. Maciejowski, J.: Predictive Control with Constraints. Prentice-Hall, Harlow (2002)
15. Mayne, D.Q., Rawlings, J.B., Rao, C.V., Scokaert, P.O.M.: Constrained model predictive control: stability and optimality. Automatica **36**, 789–814 (2000)
16. Moore, K.L.: Iterative Learning Control for Deterministic Systems. Advances in Industrial Control. Springer, London (1993)
17. Morari, M., Lee, J.H.: Model predictive control: past, present and future. Comput. Chem. Eng. **23**, 667–682 (1999)
18. Nandan, A., Imtiaz, S.: Nonlinear model predictive control of managed pressure drilling. ISA Trans. **69**, 307–314 (2017)
19. Nelles, O.: Nonlinear System Identification. From Classical Approaches to Neural Networks and Fuzzy Models. Springer, Berlin (2001)
20. Nørgaard, M., Ravn, O., Poulsen, N., Hansen, L.: Networks for Modelling and Control of Dynamic Systems. Springer, London (2000)
21. Patan, K.: Approximation of state-space trajectories by locally recurrent globally feed-forward neural networks. Neural Netw. **21**, 59–63 (2008)
22. Patan, K.: Artificial Neural Networks for the Modelling and Fault Diagnosis of Technical Processes. Lecture Notes in Control and Information Sciences. Springer, Berlin (2008)
23. Scokaert, P., Clarke, D.W.: Stabilizing properties of constrained predictive control. IEE Proc. Control Theory Appl. **141**(5), 295–304 (1994)
24. Skogestad, S., Postlethwaite, I.: Multivariable Feedback Control. Analysis and Design, 2nd edn. Wiley, New York (2005)
25. Sridhar, A., Govindarajan, A., Rhinehart, R.R.: Demonstration of leapfrogging for implementing nonlinear model predictive control on a heat exchanger. ISA Trans. **60**, 218–227 (2016)
26. Tatjewski, P.: Advanced Control of Industrial Processes. Springer, London (2007)
27. Tornil-Sin, S., Ocampo-Martinez, C., Puig, V., Escobet, T.: Robust fault detection of nonlinear systems using set-membership state estimation based on constraint satisfaction. Eng. Appl. Artif. Intell. **25**(1), 1–10 (2012)
28. Verron, S., Tiplica, T., Kobi, A.: Fault diagnosis of industrial systems by conditional gaussian network including a distance rejection criterion. Eng. Appl. Artif. Intell. **23**(7), 1229–1235 (2010)
29. Xu, J.X., Tan, Y.: Linear and Nonlinear Iterative Learning Control for Deterministic Systems. Lecture Notes in Control and Information Sciences, vol. 291. Springer, Berlin (2003)

Chapter 2
Neural Networks

2.1 Introduction

A neural network, also sometimes referred to as an *artificial* neural network, is a kind of processing structure the behaviour of which is inspired by the way human brains work. A neural network story started in 1943 when a neurophysiologist Warren McCulloch and a mathematician Walter Pitts published their paper on modelling simple neural networks using electrical circuits [63]. Since then we have been witnessing a rapid development of neural network technology. According to Maass [60], the models of neural networks can be classified into the following generations:

The first generation (1940s/1950s) – this generation of neural models is based on McCulloch–Pitts threshold neurons as computational units. Neural networks consisting threshold units are referred to as perceptrons [90]. Such models are characterised by synchronous inputs and a binary (digital) output. The first generation of neural networks include a variety of structures, such as multilayer perceptrons, Hopfield networks [39] or Boltzmann machines [1]. The neural models that belong here are useful when dealing with digital data and can process any Boolean function. There are neat learning algorithms for adjusting the weights. However, perceptrons are fundamentally limited in what they can learn to do.

The second generation (1980s) – the second generation of neural networks includes models using a continuous activation function acting over the weighted sum (or polynomial) of inputs. The following types of activation functions can be distinguished: squashing functions (sigmoid or hyperbolic tangent), linear saturated function, polynomials, piecewise exponential or Gaussian functions. The typical examples of neural models of the second generation include feed-forward networks, recurrent networks and radial basis function (RBF) networks [34, 35]. The main characteristic of these models is the ability to process analog data. These

Portions of the chapter reused by permission from Springer Nature, Artificial Neural Networks for the Modelling and Fault Diagnosis of Technical Processes by Krzysztof Patan ©2008.

© Springer Nature Switzerland AG 2019

K. Patan, *Robust and Fault-Tolerant Control*, Studies in Systems, Decision and Control 197, https://doi.org/10.1007/978-3-030-11869-3_2

models are able to approximate any nonlinear mapping with arbitrary accuracy using suitable network configuration [15, 25, 40, 46, 57, 71]. As a consequence of the fact that continuous activation functions are being used in this context, numerous learning algorithms have been developed with the flagship example being the so-called back-propagation (BP) algorithm [92].

The third generation (1990s) – the third generation of neural networks was motivated by the fact that biological neural systems use timing of single action potentials, or *spikes*, to encode information [60]. Within this generation one can find models that employ spiking neurons as computational units. Each neuron incorporates the timing of individual spikes, with time playing a central role and increasing the resemblance to its biological counterpart in terms of neural simulation. Therefore, in addition to a neuronal and a synaptic state, the spiking neural networks incorporate the concept of time into their operating model [59].

Artificial neural networks have been intensively studied in the context of control theory during the past decades resulting in a number of successful application in the terms of dynamic system modelling [30, 34, 67, 68, 71, 83] as well as control [41, 55, 66, 68, 83]. Neural networks provide an interesting and valuable alternative to classical methods, since they allow one to deal with some of the most complex situations that are not handled well by deterministic algorithms that require more precise definitions. The notion of neural network is especially useful in situations when there is no mathematical model of the process considered, so the classical approaches such as observers or parameter estimation methods cannot be applied. Neural networks provide an excellent mathematical tool for dealing with nonlinear problems [15, 25, 40, 46, 57, 71]. They have an important property, namely that any nonlinear function can be approximated with arbitrary accuracy using a neural network with a suitable architecture and weight parameters. Neural networks are parallel data processing tools capable of learning functional dependencies of data. This feature is extremely useful when solving different pattern recognition problems. Yet another important property of such networks is a self-learning capability. A neural network can extract the system features from historical training data using the learning algorithm, requiring little or no a priori knowledge about the process. This allows the modelling of nonlinear systems with greater flexibility. These features allow one to design adaptive control systems for complex, unknown and nonlinear dynamic processes. Neural networks are also robust with respect to incorrect or missing data. Protective relaying based on artificial neural networks is not affected by a change in the operating conditions of a system. Moreover, neural networks are characterized by high computation rates, large input-error tolerance and adaptability in terms of the changing work conditions.

All the mentioned features means that neural networks have become very useful and popular processing tools in control applications. Taking into account the input-output relations that can be realized by neural networks, one can divide the possible neural models into static and dynamic. Figure 2.1 illustrates the most frequently used neural network models.

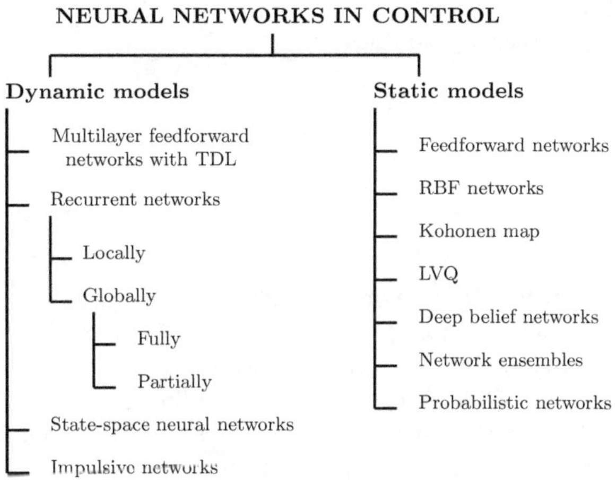

Fig. 2.1 Selected neural networks in control applications

2.2 Static Models

Static models represent the input-output mappings where the output at any given time depends on the input at the same time instant. In control theory, the static model can be used in approximating static nonlinearities, e.g. in the framework of Wiener and Hammerstein models [45, 56], in monitoring and fault diagnosis [54], in kinematics and inverse kinematics modelling [14], in realization of look-up tables [42], in approximation of control surfaces [24], etc.

2.2.1 The Model of an Artificial Neuron

Artificial neural networks are constructed using a certain number of single processing units called neurons. The McCulloch–Pitts model is the fundamental, classical neuron model and it is described by the equation:

$$y = \sigma(\varphi) = \sigma\left(\sum_{i=1}^{n} w_i u_i + b\right), \tag{2.1}$$

where u_i, $i = 1, 2, \ldots, n$, denote neuron inputs, b is the bias (threshold), w_i denotes synaptic weight coefficients, φ is the activation signal, $\sigma(\cdot)$ is a nonlinear activation function. The bias parameters can be implemented in the form of the additional input weight excited by the constant signal of the value equal to 1. The input-output representation of the artificial neuron is shown in Fig. 2.2. In their original proposition, McCulloch and Pitts were only able to represent binary functions as the neuron was equipped with a simple thresholding function. However, the importance of this

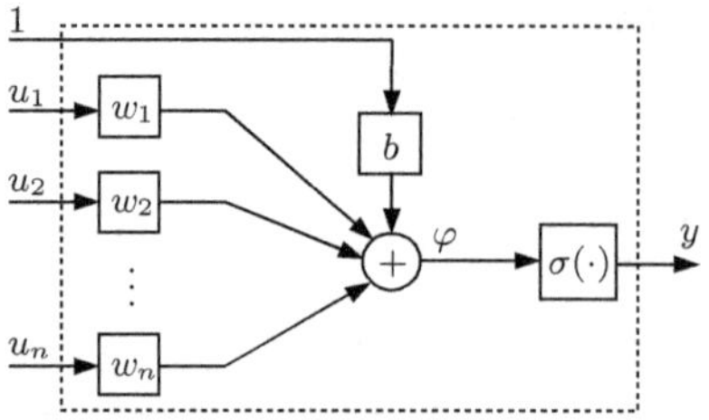

Fig. 2.2 Neuron scheme with n inputs and one output

model lied in its pioneering character and providing a motivation for developing more advanced models of the neuron. The most frequently appearing modification of this initial model consists in using a more complex form of the activation function. In 1960, Widrow and Hoff applied a linear activation function and thus created the *Adaline* neuron [107, 108]. In order to make the neural network applicable to a wide class of nonlinear problems, the squashing functions are most often used [34, 35]. In some specific neural structures, certain other nonlinear activation functions can be applied, e.g. radial basis functions [27, 83], exponential functions [98], etc. The choice of a suitable activation function is dependent on a specific application of the neural network.

2.2.2 *Feed-Forward Multilayer Networks*

A multilayer perceptron is a network, in which the neurons are grouped into layers (Fig. 2.3). Such a network has an input layer, one or more hidden layers, and an output layer. The main task of the input units (black circles) is to perform preliminary input data processing $\boldsymbol{u} = [u_1, u_2, \ldots, u_n]^{\mathsf{T}}$ and to pass them to the elements of the hidden layer for further processing. Data processing can comprise scaling, filtering or signal normalization. The fundamental neural data processing is carried out in the hidden and output layers. It should be emphasised that the links between neurons are designed in a way so that each element of the previous layer is connected with each element of the next layer. These connections are given suitable weight coefficients determined, for each separate case, depending on the task the network is designed to solve. The output layer generates the network response vector $\boldsymbol{y} = [y_1, y_2, \ldots, y_m]^{\mathsf{T}}$. The nonlinear neural computing performed by the network shown in Fig. 2.3 can be expressed by:

$$\boldsymbol{y} = \boldsymbol{\sigma}_3\left(\boldsymbol{W}_3\boldsymbol{\sigma}_2\left(\boldsymbol{W}_2\boldsymbol{\sigma}_1(\boldsymbol{W}_1\boldsymbol{u})\right)\right), \tag{2.2}$$

where $\boldsymbol{\sigma}_1$, $\boldsymbol{\sigma}_2$ and $\boldsymbol{\sigma}_3$ are vector-valued activation functions defining neural signal transformation through the first, the second and the output layers; $\boldsymbol{W}_1$, $\boldsymbol{W}_2$ and $\boldsymbol{W}_3$ are the matrices of weight coefficients that determine the intensity of connections among neurons in the neighbouring layers; $\boldsymbol{u}$, $\boldsymbol{y}$ are the input and output vectors, respectively.

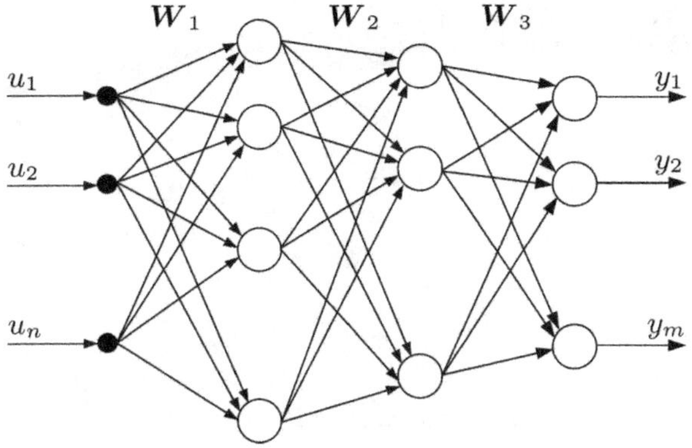

Fig. 2.3 A forward network with two hidden layers

One of the fundamental advantages of neural networks is that they have the capability of learning and adapting. From the technical point of view, the training of a neural network is nothing else but a determination of weight coefficient values among the neighbouring processing units. The back-propagation algorithm [91, 92, 106] is the fundamental training algorithm used in feed-forward multilayer networks. BP gives a procedure allowing one to change the arbitrary weight value assigned to the connection among the processing units in the neighbouring layers of the network. This algorithm is of an iterative type and it is based on the minimization of a sum-squared error utilizing gradient descent for optimization. The modification of the weights is performed according to the formula:

$$\boldsymbol{w}(k+1) = \boldsymbol{w}(k) - \eta \nabla J\left(\boldsymbol{w}(k)\right), \tag{2.3}$$

where $\boldsymbol{w}(k)$ denotes the weight vector at the iteration k, η is a learning rate, and $\nabla J\left(\boldsymbol{w}(k)\right)$ is the gradient of the performance index J with respect to the weight vector $\boldsymbol{w}$. The back-propagation algorithm is widely used, however it has a slow convergence rate and often gets stuck in local minima of the cost function. To overcome such limitations, certain modifications can be introduced in the form of e.g. the momentum [84] or adaptable learning parameters [17]. There are many other modifications of BP, which have proved their usefulness in practical applications. Naming just a few: the quickprop algorithm [21], resilient back-propagation [89], conjugate gradient methods [32], quasi-Newton methods [68], the Levenberg–Marquardt (LM) algorithm [33], stochastic learning methods [75, 76], etc.

2.2.3 Radial Basis Networks

The radial basis function network approach is becoming more popular as an alternative solution to the slowly convergent multilayer perceptrons. Similarly as in the case of multilayer perceptrons, within the radial basis network approach one can

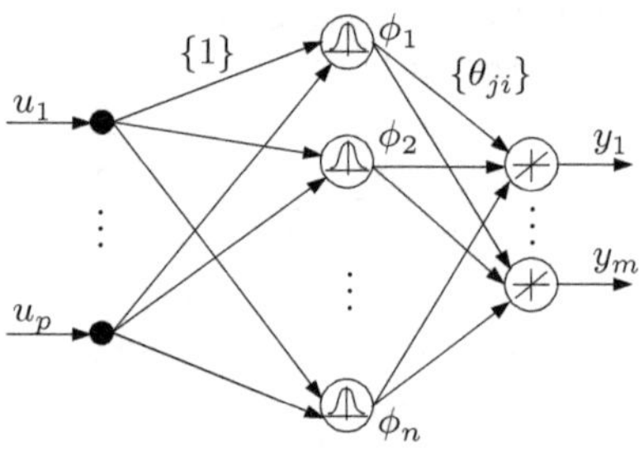

Fig. 2.4 A structure of a radial basis function network

model any nonlinear function [27]. However, this kind of a neural network requires a large number of nodes in order to achieve the required approximating properties. This choice of the number of nodes is of similar character to the choice of the number of the hidden layers and neurons in a multilayer perceptron. The RBF network architecture is shown in Fig. 2.4. Such a network has three layers: the input layer, the single (hidden) nonlinear layer and the linear output layer. It should be noticed that the weights connecting the input and the hidden layer have values equal to one. This means that input data are passed on to the hidden layer with no weight operation. The output ϕ_i of the nth neuron of the hidden layer is a nonlinear function of the Euclidean distance between the input vector $\boldsymbol{u} = [u_1, \ldots, u_p]^\mathsf{T}$ and the vector of the centres $\boldsymbol{c}_i = [c_{i1}, \ldots, c_{ip}]^\mathsf{T}$, and can be described by the following expression:

$$\phi_i = \phi\left(\|\boldsymbol{u} - \boldsymbol{c}_i\|, \rho_i\right), \quad i = 1, \ldots, n, \tag{2.4}$$

where ρ_i denotes the spread of the ith basis function, $\|\cdot\|$ — the Euclidean norm. The jth network output y_j is a weighted sum of the outputs of the hidden neurons:

$$y_j = \sum_{i=1}^{n} \theta_{ji}\phi_i, \quad j = 1, \ldots, m, \tag{2.5}$$

where θ_{ji} denotes the connecting weight between the ith hidden neuron and the jth output. Many different functions $\phi(\cdot)$ were suggested. The Gaussian functions are some of the most popular choices:

$$\phi(z, \rho) = \exp\left(-\frac{z^2}{\rho^2}\right). \tag{2.6}$$

The fundamental operation performed in a RBF network is the selection of the function number, function centres and their position. An insufficient number of the centres can result in weak approximating properties. On the other hand, the number of exact centres increases exponentially with the size of the network input space. Hence, the RBF networks are not used in problems with a large input space. In order to train a RBF network, the hybrid techniques are used. First, the centres and the widths of the basis functions are established heuristically. Then, the adjusting of the weights is performed. The centres of the radial basis functions can be chosen in many ways e.g. as values of a random distribution over the input space or by clustering algorithms [13, 105], which give statistically the best choice of the centre numbers and as well as of their positions. When the centre values are established, the objective of the learning algorithm is to determine the optimal weight matrix $\Theta = \{\theta_{ji}\}$, which minimizes the difference between the desired and the real network response. The output of the network is linear in weights therefore one can use traditional regressive methods for estimation of the weight matrix [105].

2.2.4 A Kohonen Network

A Kohonen network is the self-organizing map (SOM). Such networks can learn to detect regularities and correlations in their input and adapt their future responses to that input accordingly. The network parameters are adapted by learning procedure based on input patterns only (unsupervised learning) [34]. In contrast to the standard supervised learning methods, the unsupervised ones use input signals to extract knowledge from the data. During the learning phase, there is no feedback to environment. Therefore, neurons and weighted connections should have certain level of self-organizing properties. Two dimensional SOM is shown in Fig. 2.5. The inputs and neurons in the competitive layer are all connected with each other. Note that the competitive layer is also the output layer generating a response of a Kohonen network. The weight parameters are adapted using the *winner-takes-all* rule, in the following fashion [52]:

$$\|\boldsymbol{u} - \boldsymbol{w}_c\| = \min_i\{\|\boldsymbol{u} - \boldsymbol{w}_i\|\}, \tag{2.7}$$

where $\boldsymbol{u}$ is the input vector, $\boldsymbol{w}_c$ is the winner's weight vector and $\boldsymbol{w}_i$ is the weight vector of the ith processing unit. However, instead of adapting the winning neuron only, all neurons within a certain neighbourhood of the winner Ω are adapted giving the *winner-takes-most* formula:

$$\boldsymbol{w}_i^{new} = \boldsymbol{w}_i^{old} + \alpha S(i, u)(\boldsymbol{u} - \boldsymbol{w}_i), \tag{2.8}$$

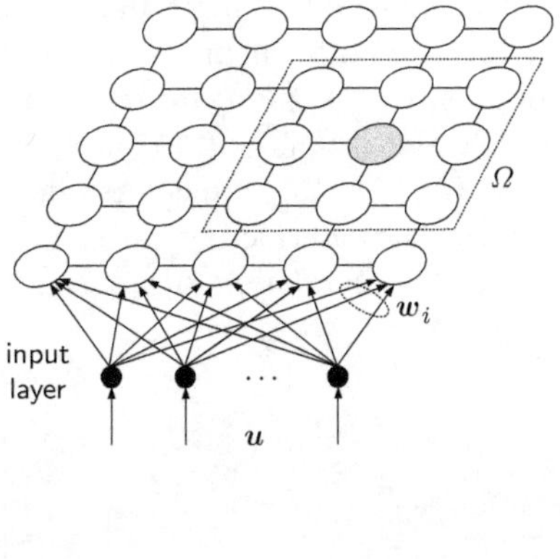

Fig. 2.5 A two-dimensional self-organizing map

where $\boldsymbol{w}_i^{new}$, $\boldsymbol{w}_i^{old}$ are the updated and original weight vectors of the ith neuron, respectively, α represents the learning rate and $S(i, u)$ is the neighbourhood function. The learning rate and neighbourhood size are altered through two phases: an ordering phase and a tuning phase. An iterative character of the learning rate allows the network to gradually establish the feature map. During the first phase, the neuron weights are expected to order themselves in the input space consistent with the associated neuron positions. During the second phase, the learning rate continues to decrease (at a slow pace). The small value of the learning rate fine-tune the network while keeping the ordering learned in the previous phase stable. In the Kohonen learning rule, the learning rate is a monotone (decreasing) time function. The functions that are used most frequently include the following: $\alpha(t) = 1/t$ or $\alpha(t) = at^{-a}$ for $0 < a \leqslant 1$. The concept of the neighbourhood is extremely important in terms of network processing. In most cases, either a rectangular or a hexagonal grid is used. The concept of the neighbourhood is illustrated in Fig. 2.5 with the winner marked in grey. After designing the network, the next (and very important) issue to be tackled is to assign the clustering results generated by the network with the desired results of a given problem. It is necessary to determine which regions of the feature map are to be activated for an occurrence of a given class.

2.2.5 A Learning Vector Quantization

A learning vector quantization (LVQ) is a special case of an artificial neural network, which itself is a prototype-based supervised classification algorithm. The concept of LVQ is closely related to that of the self-organizing maps. However, in contrast to SOM, which is an unsupervised learning method, LVQ utilises the supervised learning approach. Moreover, again, unlike SOM, no neighbourhoods around the winning neuron are defined, hence no spatial ordering of the codebook vectors is expected to take place [53]. A LVQ network consists of three layers of neurons as presented in Fig. 2.6: an input layer, hidden competitive layers and the output layer. Each output neuron represents one category (class). The network is fully connected when it comes to the relation among the input and the hidden layers and partially connected in terms of the hidden and output layers. Each output neuron is assigned to a different cluster of hidden (competitive) neurons. Importantly, this kind of a network has the property that the weights assigned to the connections among the hidden and output neurons are constant and equal to 1. The weights associated with the connections among the input and competitive neurons are modified in the training phase. The weights of a competitive neuron form a prototype $\boldsymbol{w}_i$. All prototypes $\boldsymbol{W} = [\boldsymbol{w}_1, \ldots, \boldsymbol{w}_n]$ are defined in the feature space of the observed data. Using the winner-takes-all rule, for each data point $\boldsymbol{x}$, one determines the prototype the closest to the input in relation to a given distance measure using (2.9):

$$|\boldsymbol{w}_i^* - \boldsymbol{x}| \leq |\boldsymbol{w}_i - \boldsymbol{x}| \quad \forall\, i. \tag{2.9}$$

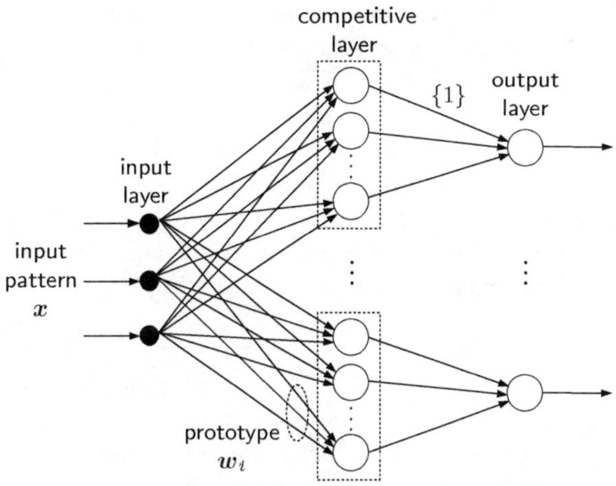

Fig. 2.6 A LVQ neural network

Then, the weights of the winner neuron $\boldsymbol{w}_i^*$ are modified. Both the hidden neurons and the output neurons have binary outputs. When an input pattern is supplied to the network, the hidden neuron the prototype of which is closest to the input pattern is said to win the competition for being activated and thus allowed to produce a value of 1. All other neurons in the competitive layer are forced to give a response equal to 0. The output neuron connected to the cluster of the competitive neurons that contains the winner also generates the value equal to 1 with all the other output neurons yielding the value of 0. The output neuron that produces the value of 1 returns the class to which the input pattern is assigned.

2.2.6 Deep-Belief Networks

Deep belief networks (DBNs) are probabilistic generative models composed of multiple layers of stochastic, latent variables (Fig. 2.7a). The latent variables typically have binary values and are called hidden units. In Fig. 2.7a the top two layers have undirected, symmetric connections among them and form an associative memory.

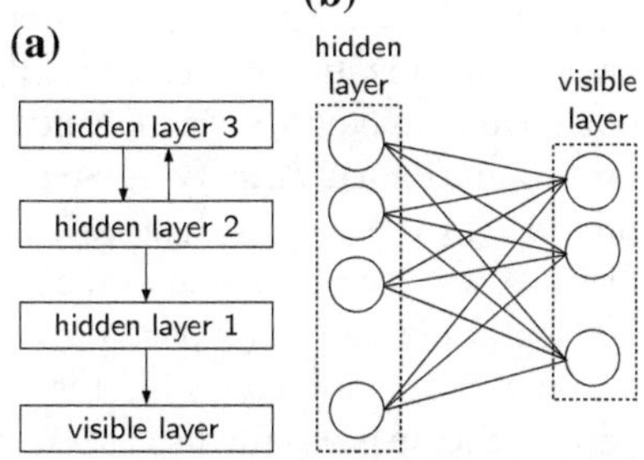

Fig. 2.7 A deep belief network (**a**) and a restricted Boltzmann machine (**b**)

The lower layers receive top-down, directed connections from the layer placed above. DBNs can be treated as a set of simple, unsupervised networks, such as restricted Boltzmann machines (RBMs) [94] where each sub-network's hidden layer serves as the visible layer for the next one. RBM is an undirected, generative energy-based model with the so-called visible input/output layer and one hidden layer. There is no connection among the units within the same layer (Fig. 2.7b). When the network is used to generate data, the probability of turning on the ith neuron is a logistic function of the states of its immediate ancestors j and of the weights, w_{ij}, on the directed connections from the ancestors [36]:

$$p(h_i = 1) = \frac{1}{1 + \exp(-b_i - \sum_j h_j w_{ij})}, \tag{2.10}$$

where b_i represents the bias of the ith neuron.

The DBN model is trained sequentially, one module after another, proceeding from the lowest RBM and using its outputs (after its training is completed) as inputs for the training of the subsequent RBM module. Each RBM is trained in an unsupervised manner. The training method proposed by Hinton is called a contrastive divergence method and provides an approximation to the maximum likelihood method. For each weight w_{ij} between the ith visible unit v_i and the jth hidden unit h_j we have:

$$\Delta w_{ij} = \eta \frac{\partial \log(p(\boldsymbol{v}))}{\partial w_{ij}}, \tag{2.11}$$

where η is the learning rate and $p(\boldsymbol{v})$ stands for the probability of the visible vector $\boldsymbol{v}$. After the training of all RBM modules is finished, the output of the final one can be used as representation of the inputs with reduced dimensionality [37].

2.2.7 A Neural Network Ensemble

In many cases, a single neural network neither ensures the required mapping nor provides the required generalization capabilities. Therefore, if one designs networks that are targeting only the specific parts of the complete mapping, it could mean that the overall task is performed more efficiently. The idea behind the neural network ensemble is to combine different neural classifiers into a meta-classifier that has a better generalization performance than each of the individual classifiers alone. The objective is to develop n independently trained neural networks playing the role of experts and then to classify a given input pattern based on a combination of responses provided by those individual experts. The final evaluation is derived using a decision block. The general scheme of the neural network ensemble is presented in Fig. 2.8.

Neural classifiers (experts) are trained simultaneously for the same task. The decision block is responsible for the final evaluation. In order to accomplish it, a voting scheme is usually used [112]. For the classification purposes, the most popular

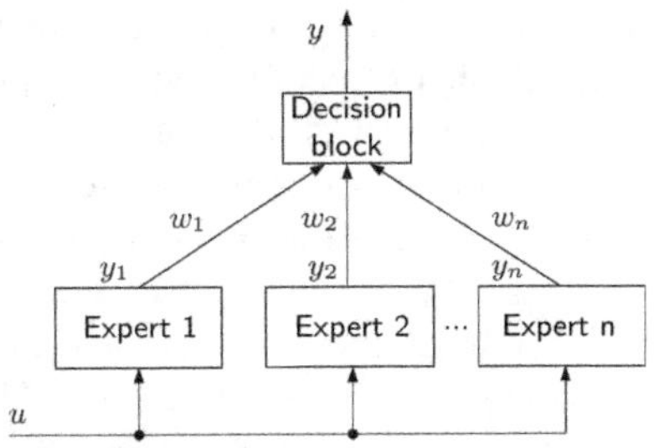

Fig. 2.8 A parallel expert scheme

methods include the plurality vote and the majority vote. Note that when the ensemble is used in regression, it is the weighted average that is frequently used (Fig. 2.8) [95]:

$$y = w_1 y_1 + w_2 y_2 + \cdots + w_n y_n, \tag{2.12}$$

where w_i are weights, and y_i are the predictions given by individual experts.

Experts can be implemented using artificial neural networks in the form of e.g. multilayer perceptrons [5] or radial basis networks [62]. Each neural network can have an optional structure and should be trained with a convenient algorithm using different learning sets. The only condition is that each neural network has to have the same number of outputs.

2.2.8 *Probabilistic Networks*

A probabilistic neural network (PNN) is a feed-forward neural network widely used in classification and pattern recognition. It was invented by Specht [98]. In PNN, the parent probability distribution function of each class is estimated by a Parzen window and a non-parametric function. Using a probability distribution function of each class, the class probability of a new input data is derived and Bayes rule is then employed to allocate the class with the highest posterior probability to new input data. PNN shown in Fig. 2.9 has a feed-forward structure with four layers. The input units are simply distribution units that supply the same input values to all of the units in

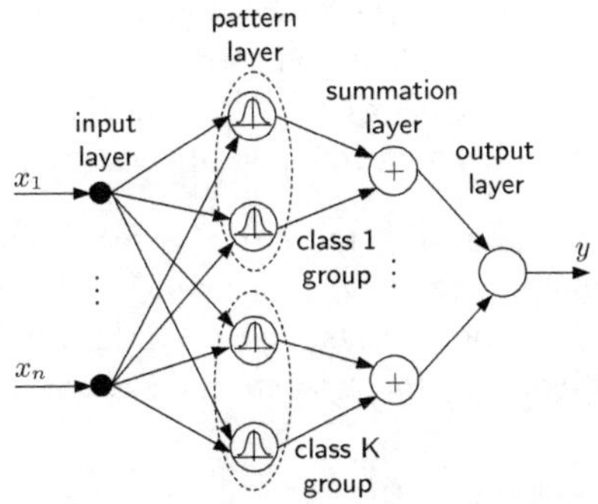

Fig. 2.9 A probabilistic neural network

the hidden layer. Each unit in the pattern layer represents a training pattern. A pattern unit calculates the dot product of the input pattern $\boldsymbol{x}$ with a weight vector $\boldsymbol{w}_i$. The calculated scalar $z_i = \boldsymbol{x} \cdot \boldsymbol{w}_i$ is then processed by a nonlinear function. The radial basis function is frequently used for that purpose. The summation neurons simply sum the inputs from the pattern units that correspond to the class from which the training pattern was selected (see Fig. 2.9). The output layer compares the weighted votes for each target category accumulated in the pattern layer and uses the largest vote to predict the target category. When no a priori information about the classes is known, the Bayesian theory is applied as follows:

$$class(\boldsymbol{x}) = \arg\max_{1 \le i \le NC} P(\boldsymbol{x}|C_i), \tag{2.13}$$

where NC is the number of classes and $P(\boldsymbol{x}|C_i)$ is the class conditional probability of class C_i. As described by Specht [98], the network is trained by setting the weight vector $\boldsymbol{w}_i$ in one of the pattern units equal to each of the $\boldsymbol{x}$ patterns in the training set and then connecting the pattern neuron output to the appropriate summation unit. The pattern neurons representing the same class are grouped (see Fig. 2.9) by separate summation units to provide additional bits of information in the output vector.

2.3 Dynamic Models

Dynamic neural networks have structures that are fundamentally different from the ones found within the static ones. Such a class of neural models makes it possible to take into account the more dynamic properties of the plant considered. Dynamics can be realized either in an external or an internal way [34, 67, 72]:

- *The external-dynamics approaches.* The most frequently utilized nonlinear dynamic system modeling and identification approach relies o an external-dynamics strategy. The idea behind the external-dynamics approach is to use a nonlinear static approximator equipped with an external-dynamics filter realized in the form of time delays of the input signals. Such structures are also known as time-delay neural networks (TDNNs);
- *The internal-dynamics approaches.* The models with internal dynamics are based on the extension of static models with an internal memory. The memory is realized by means of feedback mechanisms. Therefore, such neural structures are also called recurrent neural networks.

In contrast to the models with external dynamics, the use of past inputs and past outputs in the model input is not necessary. Therefore, the application of internal dynamic models leads to a desirable reduction of the input space dimensionality. Such dynamic systems have powerful modelling capabilities that have not been fully explored so far. From the point of view of a potential feedback, the recurrent networks can be divided as follows:

- Local recurrent networks — with feedbacks only inside neuron models. Such networks have a structure similar to the static ones but consisting of dynamic neuron models.
- Global recurrent networks — with feedbacks allowed among neurons of different layers or among neurons of the same layer. In the particular case, when feedbacks link the output and the input layers, the network is called a recurrent network with outer feedback.

2.3.1 *Neural Networks with External Dynamics*

The multilayer perceptron is the most commonly used neural network in terms of modelling. This class of neural models, however, is of a static type and can only be used to approximate a continuous nonlinear, although static, function [15, 40]. Therefore, the neural network modelling of control systems should take into account the dynamics of processes or systems considered. There are two main methods providing a static neural network with dynamic properties: the insertion of an external memory to the network or the use of feedback. The strategy most frequently applied to model dynamic nonlinear mapping is the external dynamics approach [34, 41, 66–68]. It is based on the nonlinear input/output model as in (2.14):

$$\hat{y}(k+1) = f\big(y(k), \ldots, y(k-n_a+1), u(k), \ldots, u(k-n_b+1)\big), \tag{2.14}$$

where $f(\cdot)$ is a nonlinear function, $u(k)$ is the input, $y(k)$ and $\hat{y}(k)$ are outputs of the process and the model, respectively, n_a and n_b represent the number of delayed output and input signals, respectively. The nonlinear model is clearly separated into two parts: a nonlinear static approximator (multilayer perceptron) and an external dynamic filter bank (tapped delay lines (TDL)) (Fig. 2.10). As a result, a model known

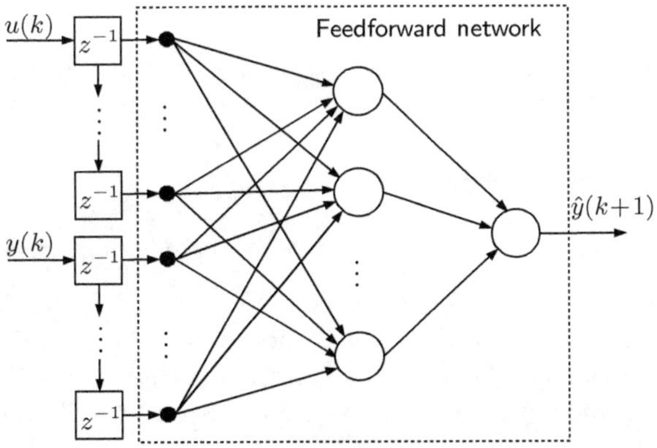

Fig. 2.10 External dynamics approach realization

as a multilayer perceptron with tapped delay lines (a time-delay neural network) is obtained. The notion of time-delay neural network can be applied to a large class of systems but are as general as the nonlinear state-space models. The limitations of such an approach can be observed for processes with non-unique nonlinearities, e.g. hysteresis or backslash, where the internal unmeasurable states play a decisive role, and partly for processes with non-invertible nonlinearities [67, 113]. Moreover, the problem of order selection is not satisfactorily solved yet. This problem is equivalent to the determination of relevant inputs for the function $f(\cdot)$. If the order of a process is known, all the necessary past inputs and outputs should be fed into the network. This way, the input space of the network grows large. In many practical cases, there is no possibility to learn the order of the modelled process, and the number of suitable delays has to be selected experimentally using the trial-and-error procedure [68].

In general, nonlinear dynamic models with external dynamics can be expressed in the following form:

$$\hat{y}(k+1) = f(\varphi(k)), \tag{2.15}$$

where $\varphi(k)$ is the regression vector. The form of the regression vector determines the type of the neural model [67, 68]:

- nonlinear auto-regressive with exogenous inputs model (NARX):

$$\varphi(k) = [u(k), \ldots, u(k-n_b+1), y(k), \ldots, y(k-n_a+1)]^{\mathsf{T}},$$

- nonlinear auto-regressive model (NAR):

$$\varphi(k) = [y(k), \ldots, y(k-n_a+1)]^{\mathsf{T}},$$

- nonlinear finite impulse response model (NFIR):

$$\varphi(k) = [u(k), \ldots, u(k-n_b+1)]^{\mathsf{T}},$$

- nonlinear output error model (NOE):

$$\varphi(k) = [u(k), \ldots, u(k-n_b+1), \hat{y}(k), \ldots, \hat{y}(k-n_a+1)]^{\mathsf{T}},$$

- nonlinear auto-regressive moving average with exogenous inputs model (NARMAX):

$$\varphi(k)=[u(k), \ldots, u(k-n_b+1), y(k), \ldots, y(k-n_a+1), e(k), \ldots, e(k-n_c+1)]^{\mathsf{T}},$$

  where n_c is the number of delayed prediction errors $e(k) = y(k) - \hat{y}(k)$.

The NOE and NARMAX models require the knowledge of the previous model outputs and possibly of the prediction errors. These values are provided by means of the output feedback mechanism. Such models are trained using a parallel identification scheme [66], described in detail in Sect. 2.4.1. The main drawback of the models with the output feedback is that in such cases assuring stability can be problematic. Another problem is that the choice of the model order is crucial for the performance and no really efficient methods for its determination are available [67]. The use of real plant outputs avoids many of the analytical difficulties encountered, assures stability and simplifies the identification procedure. These models, e.g. NARX or NAR are trained using a series-parallel model [66] discussed later on in Sect. 2.4.1. Such networks are capable of modelling systems, if they have a weakly visible state, i.e. if there is an input-output equivalent to a system whose state is a function or a fixed set of finitely many past values of its inputs and outputs. Otherwise, the model has a strongly hidden state and its identification requires recurrent networks of a fairly general type.

2.3.2 Fully Recurrent Networks

The most general architecture of a recurrent neural network was proposed by Williams and Zipser [111]. This structure is often called the real-time recurrent network (RTRN), since it was designed for the real-time signals processing. The network consists of M neurons, each of them providing feedback. All types of connections among neurons are allowed, so a fully connected neural architecture is obtained. Note that it is only m out of M neurons that are identified as the output neurons. The remaining $H = M - m$ units are the hidden ones. The scheme of the RTRN architecture with n inputs and m outputs is shown in Fig. 2.11. Let us assume that $\boldsymbol{u}(k)$ denotes the input external vector at time k, and $\boldsymbol{y}(k+1)$ is the output vector all of the neurons at time $k+1$. The vectors $\boldsymbol{u}(k)$ and $\boldsymbol{y}(k)$ compose the forcing vector $\boldsymbol{v}(k)$ for the neurons:

$$\boldsymbol{v}(k) = [u_1(k), \ldots, u_n(k), y_1(k), \ldots, y_m(k)]^{\mathsf{T}} . \tag{2.16}$$

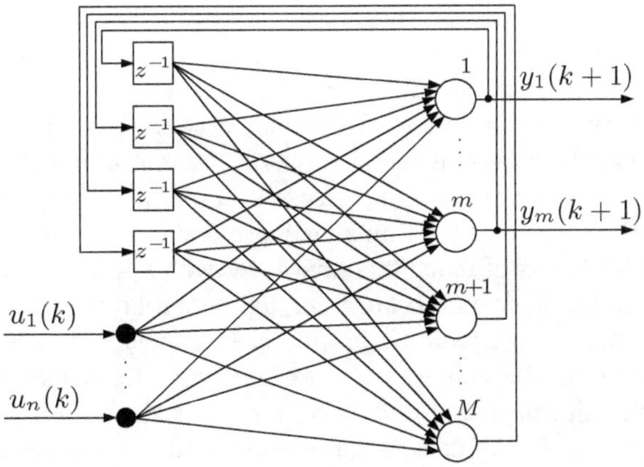

Fig. 2.11 A fully recurrent network of Williams and Zipser

Let us assume that $\boldsymbol{W}$ denotes the $M \times (M+n)$ weight matrix. The weighted sum of the inputs to the ith neuron at time k is defined according to the formula:

$$x_i(k) = \sum_{j=1}^{M+n} w_{ij}(k)v_j(k). \tag{2.17}$$

At the next time step, the neuron output is calculated as follows:

$$y_i(k+1) = f(x_i(k)), \tag{2.18}$$

where $f(\cdot)$ denotes a continuous activation function. Williams and Zipser developed a real-time recurrent learning (RTRL) algorithm for this kind of a neural network [111]. They found that the learning process converges, if a learning rate is sufficiently small. Many examples show that the RTRL method has a good modelling and generalization capabilities [110]. However, the memory and time requirements are very high. In each step of the algorithm, an update of $M^2 \times (M+n)$ partial derivatives for M fully connected units needs to take place. In the literature, one also finds some other learning algorithms being used, such as back-propagation through time (BPTT) [91] or time-dependent recurrent back-propagation (TDRBP) [82]. However, such algorithms all have limited applicability. The BPTT algorithm requires computers with powerful numeric and memory capabilities. Thus, this approach is impractical for long sequences or sequences of an unknown length. In turn, the TDRBP method cannot be utilised in real time, whereas it is very useful for off-line training.

2.3.3 Partially Recurrent Networks

In contrast to a fully recurrent network, the architecture of a partially recurrent network is based on a feed-forward multilayer perceptron consisting of an additional layer of units called the context layer [20, 51, 65]. The neurons of this layer serve as internal states of the model. Among many proposed structures, two partially recurrent networks have received considerable attention, namely the so-called Elman [20] and Jordan [51] structures. The Elman network is probably the best-known example of a partially recurrent neural network. The realization of such a network is considerably less expensive than in the case of a multilayer perceptron with tapped delay lines. The scheme of the Elman network is shown in Fig. 2.12a. This network consists of four layers of units: the input layer with n units, the context layer with v units, the hidden layer with v units and the output layer with m units. The input and output units interact with the outside environment, whereas the hidden and context units do not. The context units are used only to memorize the previous activations of the hidden neurons. A very important assumption is that in the Elman structure the number of context units is equal to the number of hidden units. All the feed-forward connections are adjustable; the recurrent connections denoted by the thick

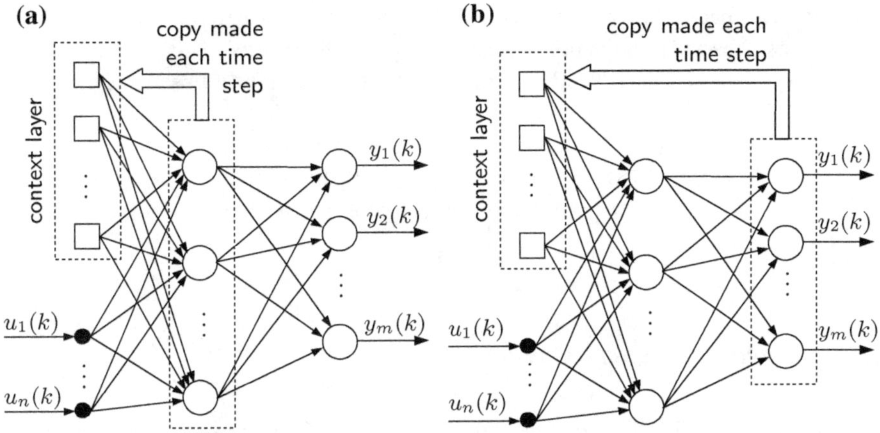

Fig. 2.12 Partially recurrent networks due to Elman (**a**) and Jordan (**b**)

arrow in Fig. 2.12a are fixed. Theoretically, this kind of networks is able to model the νth-order dynamic system, if it can be trained to do so [34]. At a specific time k, the previous activation of the hidden units (at time $k-1$) and the current inputs (at time k) are used as inputs to the network. In this case, the Elman network's behaviour is analogous to that of a feed-forward network. Therefore, the standard back-propagation algorithm can be applied to train the network parameters. However, it should be kept in mind that such simplifications limit the application of the Elman structure to the modelling of dynamic processes [35]. In turn, the Jordan network, presented in Fig. 2.12b, feedback connections from the output neurons are fed to the context units. The Jordan network has been successfully applied to recognize and differentiate various output time-sequences [50, 51] or to classify English syllables [2]. The advantage of the partially recurrent networks over the fully recurrent ones is that their recurrent links are more structured, which leads to faster training and fewer stability problems [34, 67]. Nevertheless, the number of states is still strongly related to the number of the hidden (for Elman) or the output (for Jordan) neurons, which severely restricts their flexibility. In the literature, there are propositions aimed to extend the partially recurrent networks by introducing additional recurrent links, represented by the weight α, from the context units to themselves [83, 99]. The value of α should be less than 1. For the α close to 1, the long term memory can be obtained but α becomes then less sensitive to details.

Yet another approach to the architecture can be found in the form of a recurrent network as described by Parlos [69]. A recurrent multilayer-layer perceptron (RMLP) is designed based on a multilayer perceptron network and by adding delayed links among neighbouring units of the same hidden layer (cross-talk links), including unit feedback on itself (recurrent links) [69]. Empirical evidence indicates that by using delayed recurrent and cross-talk weights, the RMLP network is able to emulate a large class of nonlinear dynamic systems. The feed-forward part of the network still maintains the well-known curve-fitting properties of the multilayer perceptron, while the feedback part provides its dynamic character. Moreover, the usage of the past process observations is not necessary, because their effect is captured by internal network states. The RMLP network has been successfully used as a model for dynamic system identification [69]. However, a drawback of this dynamic structure is increased network complexity strictly dependent on the number of hidden neurons and the resulting long training time. For the network containing one input, one output and only one hidden layer with v neurons, the number of the network parameters is equal to $v^2 + 3v$.

2.3.4 Locally Recurrent Networks

A biological neural cell not only contains a nonlinear mapping operation on the weighted sum of its inputs but it also can be said to incorporate certain dynamic properties, such as state feedbacks, time delays hysteresis or limit cycles. In order to model such a dynamic behaviour, a special kind of neuron models has been proposed [7, 22, 28, 31, 85]. Such neuron models constitute basic building blocks of a more complex dynamic neural network.

The dynamic neuron unit described by Gupta and colleagues in [30] as the basic element of neural networks of the dynamic type is presented in Fig. 2.13a. The neuron receives not only external inputs but also state feedback signals from itself and other neurons in the network. The synaptic links in this model contain a self-recurrent connection representing a weighted feedback signal of its state and lateral connections, which constitute state feedback from other neurons of the network. The dynamic neuron unit is connected to the other $(n - 1)$ models of the same type forming a neural network (Fig. 2.13b).

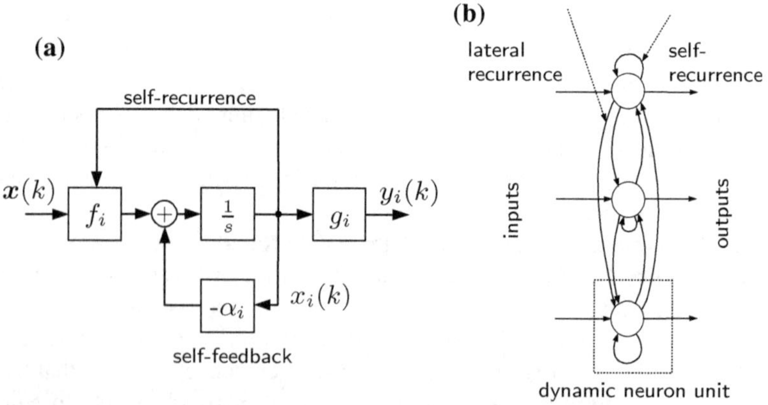

Fig. 2.13 A generalized structure of the dynamic neuron unit (**a**), network composed of dynamic neural units (**b**)

The general discrete-time representation of the ith dynamic neuron unit is as follows:

$$x_i(k+1) = -(\alpha_i - 1)x_i(k) + f_i(\boldsymbol{w}_i, \boldsymbol{x}(k)), \tag{2.19}$$

$$y_i(k) = g_i(x_i(k)), \tag{2.20}$$

where $\boldsymbol{x} \in \mathbb{R}^{n+1}$ is the augmented vector of n-neural states from other neurons in the network including the bias, $\boldsymbol{w}_i$ is the vector of synaptic weights associated with the ith dynamic neuron unit, α_i is the feedback parameter of the ith dynamic unit, $y_i(t)$ is the output of the ith neuron, $f_i(\cdot)$ is a nonlinear function of the ith neuron, and $g_i(\cdot)$ is an output function of the ith neuron. Due to various choices of the functions $f_i(\cdot)$ and $g_i(\cdot)$ in (2.19) and (2.20) as well as different types of synaptic connections, different dynamic neuron models can be obtained. Exhausting review of possible models can be found in [30].

The neural networks composed of dynamic neuron units have a recurrent structure with lateral links among neurons, as depicted in Fig. 2.13b. A different approach providing dynamically-driven neural networks is used in the so-called locally recurrent globally feed-forward (LRGF) networks [12, 71, 100]. LRGF networks have an architecture that is a halfway between the feed-forward and the globally recurrent ones. The topology of such neural networks is analogous to the multilayered feed-forward ones, and the dynamics are reproduced by the so-called dynamic neuron models. On the basis of the well-known McCulloch–Pitts neuron model, different dynamic neuron models can be designed. In general, differences among these depend on the localization of the internal feedbacks.

A model with local activation feedback. The following neuron model was studied by Frasconi [22]; it can be described by the following equations:

$$\varphi(k) = \sum_{i=1}^{n} w_i u_i(k) + \sum_{i=1}^{r} d_i \varphi(k-i), \tag{2.21a}$$

$$y(k) = \sigma\Big(\varphi(k)\Big), \tag{2.21b}$$

where $u_i(k)$, $i = 1, 2, \ldots, n$ are the inputs to the neuron, w_i reflects the input weights, $\varphi(k)$ is the activation potential, d_i, $i = 1, 2, \ldots, r$ are the coefficients which determine feedback intensity of $\varphi(k-i)$, and $\sigma(\cdot)$ is a nonlinear activation function. Viewing Fig. 2.14 as a reference point, note that the input to the neuron can be a combination of input variables and delayed versions of the activation $\varphi(k)$. Note also that the right-hand side summation in (2.21a) can be interpreted as the finite impulse response (FIR) filter. This neuron model has the feedback signal realized before the nonlinear activation block (Fig. 2.14).

A model with local synapse feedback. Back and Tsoi [7] introduced a neuron architecture with local synapse feedback (Fig. 2.15). In this structure, instead of a synapse in the form of a weight, a synapse with a linear transfer function (the Infinite

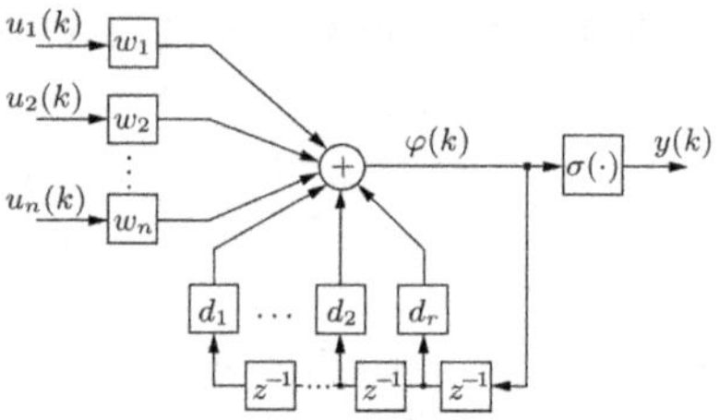

Fig. 2.14 A neuron architecture with local activation feedback

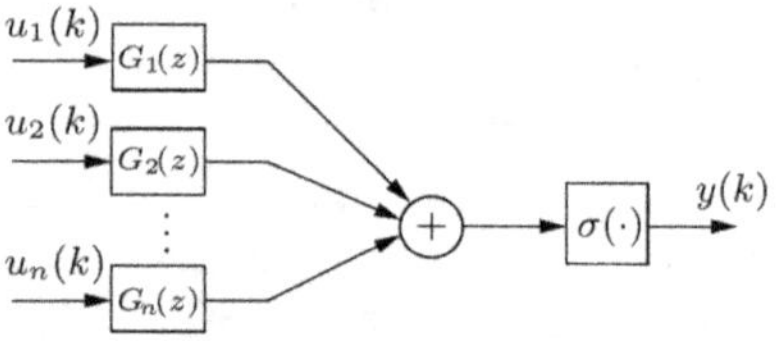

Fig. 2.15 A neuron architecture with local synapse feedback

Impulse Response (IIR) filter with poles and zeros) is used. In such a case, a neuron is described by the following set of equations:

$$y(k) = \sigma\left(\sum_{i=1}^{n} G_i(z^{-1})u_i(k)\right), \tag{2.22a}$$

$$G_i(z^{-1}) = \frac{\sum_{j=0}^{r} b_j z^{-j}}{\sum_{j=0}^{p} a_j z^{-j}}, \tag{2.22b}$$

where $u_i(k), i = 1, 2, \ldots, n$ is the set of inputs to the neuron, $G_i(z^{-1})$ is the linear transfer function, b_j, $j = 0, 1, \ldots, r$, and a_j, $j = 0, 1, \ldots, p$ are its zeros and poles, respectively. As seen in (2.22b), the linear transfer function has r zeros and p poles. Note that the inputs $u_i(k), i = 1, 2, \ldots, n$ can be taken from the outputs of the previous layer or from the output of the neuron. If these are derived from the previous layer, we end up with a local synapse feedback. If, however, these are derived from the output $y(k)$, we describe it as a local output feedback. Moreover, the local activation feedback is a special case of the local synapse feedback architecture. In this case, all synaptic transfer functions have the same denominator and only one zero, i.e. $b_j = 0, j = 1, 2, \ldots, r$.

A model with local output feedback. Another dynamic neuron architecture was proposed by Gori [28] (see Fig. 2.16). In contrast to the local synapse as well as local activation feedback approaches, this neuron model realizes feedback only after the nonlinear activation block is realized. In general, such a model can be described as follows:

$$y(k) = \sigma\left(\sum_{i=1}^{n} w_i u_i(k) + \sum_{i=1}^{r} d_i y(k-i)\right), \tag{2.23}$$

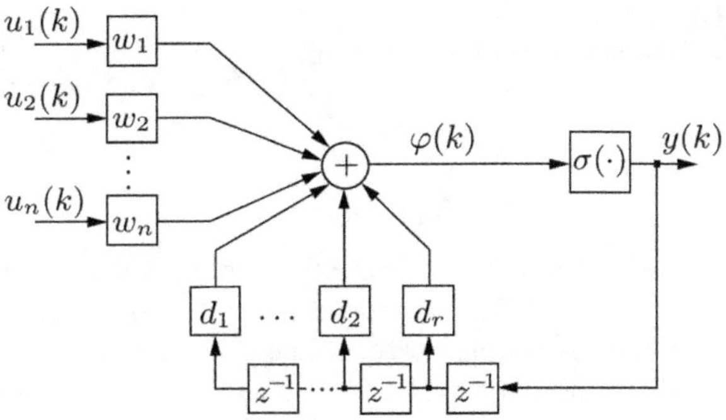

Fig. 2.16 A neuron architecture with local output feedback

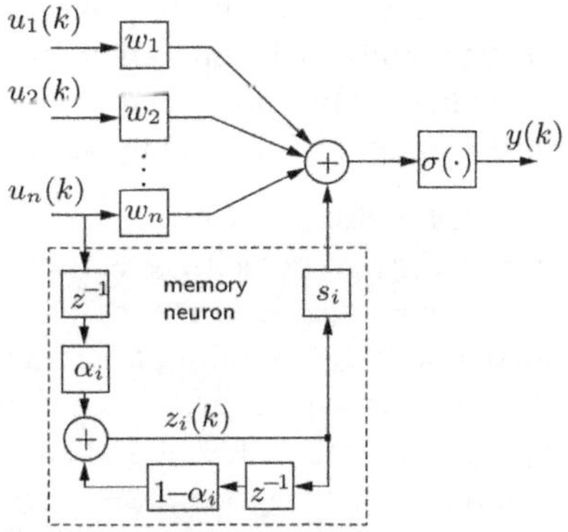

Fig. 2.17 Memory neuron architecture

where $d_i, i = 1, 2, \ldots, r$ are the coefficients that determine feedback intensity of the neuron output $y(k-i)$. In this type of architecture, the output of the neuron is filtered by the FIR filter, whose output is added to the inputs, providing the activation. It is easy to see that by the application of the IIR filter to filtering the neuron output, a more general structure can be obtained [100]. The work of Gori [28] built on the work by Mozer [65]: in fact, one can view this architecture a generalization of the Jordan–Elman architecture [20, 51].

A memory neuron. The memory neuron networks were introduced by Poddar and Unnikrishnan [85]. These networks consist of neurons that have a memory; i.e. contain information regarding past activations of its parent network neurons. A general scheme of a memory neuron is shown in Fig. 2.17.

The mathematical description of such a neuron is presented below:

$$y(k) = \sigma\left(\sum_{i=1}^{n} w_i u_i(k) + \sum_{i=1}^{n} s_i z_i(k)\right), \tag{2.24a}$$

$$z_i(k) = \alpha_i u_i(k-1) + (1-\alpha_i) z_i(k-1), \tag{2.24b}$$

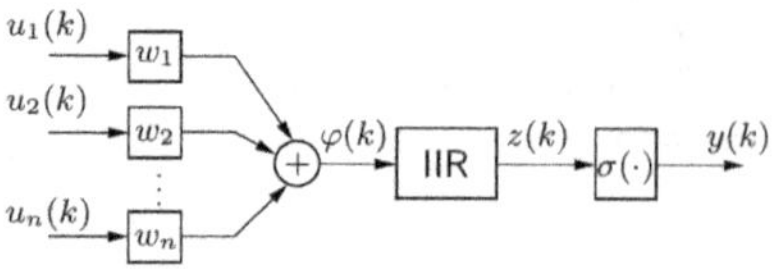

Fig. 2.18 A neuron architecture with the IIR filter

where $z_i, i = 1, 2, \ldots, n$ are the outputs of the memory neuron from the previous layer, $s_i, i = 1, 2, \ldots, n$ are the weight parameters of the memory neuron output $z_i(k)$, and $\alpha_i = \text{const}$ is a coefficient. It is observed that the memory neuron "remembers" the past output values to that particular neuron. In this case, the memory is taken to be in the form of an exponential filter. This neuron structure can be considered to be a special case of the generalized local output feedback architecture. It has a feedback-transfer function with one pole only. The memory neuron networks have been intensively studied and there exist a number of interesting results concerning the use of this architecture in the identification and control of dynamic systems [93].

A model with an IIR filter. We focus on the general structure of the neuron model proposed by Ayoubi [6] and further developed by Patan [72]. The dynamics are introduced to the neuron in such a way so that the neuron activation depends on its internal states. This is done by introducing an IIR filter into the neuron structure. This way, the neuron reproduces its own past inputs and activations using two signals: the input $u_i(k)$, for $i = 1, 2, \ldots, n$ and the output $y(k)$. Figure 2.18 shows the structure of the described neuron model. There are three main operations that are performed in this dynamic structure. First of all, the weighted sum of inputs is calculated according to the formula:

$$\varphi(k) = \sum_{i=1}^{n} w_i u_i(k). \tag{2.25}$$

The weights perform a similar role as in static feed-forward networks. The weights together with the activation function are responsible for approximation properties of the model. Then, this calculated sum $\varphi(k)$ is passed to the IIR filter. The filters are linear dynamic systems of different orders, viz. the first or the second order. The filter consists of the feedback and the feed-forward paths weighted by the weights $a_i, i = 1, 2, \ldots, r$ and $b_i, i = 0, 1, \ldots, r$, respectively. The behaviour of this linear system can be described by the following difference equation:

$$z(k) = \sum_{i=0}^{r} b_i \varphi(k-i) - \sum_{i=1}^{r} a_i z(k-i), \tag{2.26}$$

where $\varphi(k)$ is the filter input, $z(k)$ is the filter output, and k is the discrete-time index. Alternatively, the Eq. (2.26) can by rewritten in the form of the following transfer function:

$$G(z) = \frac{\sum_{i=0}^{n} b_i z^i}{1 + \sum_{i=1}^{n} a_i z^i}. \tag{2.27}$$

Finally, the neuron output can be described by:

$$y(k) = \sigma\big(g_2(z(k) - g_1)\big), \tag{2.28}$$

where $\sigma(\cdot)$ is a nonlinear activation function that produces the neuron output $y(k)$, g_1 and g_2 are the bias and the slope parameters of the activation function, respectively. In the dynamic neuron, the slope parameter can change. Thus, the dynamic neuron can model its biological counterpart to a greater degree of accuracy. In the biological neuron, at the macroscopic level, the dendrites of each neuron receive pulses at the synapses and convert them to a continuously variable dendritic current. The flow of this current through the axon membrane modulates the axonal firing rate. This morphological change of the neuron during the learning process may be modelled by introducing the slope of the activation function in the neuron as one of its adaptable parameters in addition to the synaptic weights and filter parameters [31, 114].

2.3.5 State-Space Neural Networks

Figure 2.19 shows yet another type of recurrent neural networks known as the state-space neural network (SSNN) [34, 67, 113]. The output of the hidden layer is fed back into the input layer through a bank of unit delays. The number of unit delays used here determines the order of the system. The user can choose how many neurons are used to produce feedback. Let $\boldsymbol{u}(k) \in \mathbb{R}^p$ be the input vector, $\boldsymbol{x}(k) \in \mathbb{R}^n$ — the output of the hidden layer at time k, and $\boldsymbol{y}(k) \in \mathbb{R}^m$ — the output vector. Then, the state-space representation of the neural model presented in Fig. 2.19 can be described by the equations:

$$\boldsymbol{x}(k+1) = \boldsymbol{f}\left(\boldsymbol{x}(k), \boldsymbol{u}(k)\right), \tag{2.29}$$

$$\boldsymbol{y}(k) = \boldsymbol{C}\boldsymbol{x}(k), \tag{2.30}$$

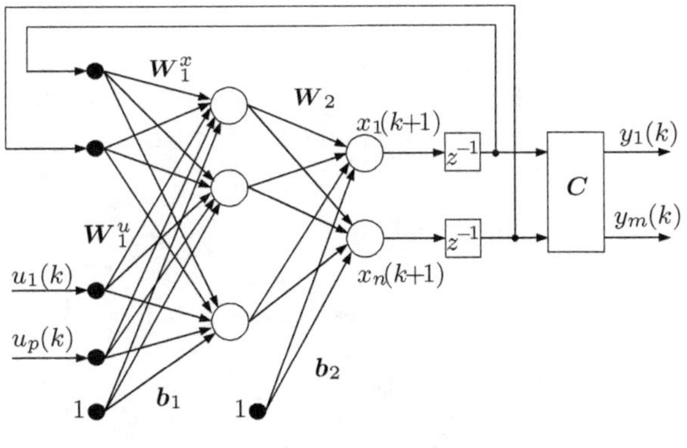

Fig. 2.19 A structure of the state-space neural network

where $\boldsymbol{f}(\cdot,\cdot)$ is a nonlinear function represented as:

$$\boldsymbol{f}(\boldsymbol{x}(k),\boldsymbol{u}(k)) = \boldsymbol{W}_2\boldsymbol{\sigma}(\boldsymbol{W}_1^x\boldsymbol{x}(k) + \boldsymbol{W}_1^u\boldsymbol{u}(k) + \boldsymbol{b}_1) + \boldsymbol{b}_2), \tag{2.31}$$

where $\boldsymbol{W}_1^x \in \mathbb{R}^{v\times n}$, $\boldsymbol{W}_1^u \in \mathbb{R}^{v\times p}$ $\boldsymbol{W}_2 \in \mathbb{R}^{n\times v}$ are input-to-hidden, state-to-hidden and hidden-to-output layers weight matrices, respectively; $\boldsymbol{b}_1 \in \mathbb{R}^v$ and $\boldsymbol{b}_2 \in \mathbb{R}^n$ are bias vectors; $\boldsymbol{\sigma} : \mathbb{R}^v \to \mathbb{R}^v$ is the vector-valued activation function of the hidden layer and v stands for the number of hidden neurons. Finally $\boldsymbol{C} \in \mathbb{R}^{m\times n}$ stands for the output (observation) matrix. It is assumed here that in order to realize the function $\boldsymbol{f}(\cdot,\cdot)$, the units of the second layer are defined using a linear activation function. This model looks similar to the external dynamic approach presented in Fig. 2.10 but the main difference is that for the external dynamics the outputs that are fed back are known during the training phase, while for the state-space model these remain unknown. As a result, state-space models can be trained only by minimizing the simulation error. The structure depicted in Fig. 2.19 and represented by (2.30) and (2.31) can be easily extended to the nonlinear state-space innovation form (SSIF) [68, 97]. The main idea is to introduce a prediction error feedback to the network structure as follows:

$$\boldsymbol{x}(k+1) = \boldsymbol{f}\left(\boldsymbol{x}(k),\boldsymbol{u}(k),\boldsymbol{e}(k)\right), \tag{2.32}$$

$$\boldsymbol{y}(k) = \boldsymbol{C}\boldsymbol{x}(k), \tag{2.33}$$

where the prediction error is defined as $\boldsymbol{e}(k) = \boldsymbol{y}_m(k) - \boldsymbol{y}(k)$, and $\boldsymbol{y}_m(k)$ stands for the output of the modelled process. As pointed out in [97], the neural network state-space innovation form can be regarded as an extended Kalman filter. Therefore, such kinds of neural models can be especially useful when dealing with observer-based control system design for nonlinear plants.

Often, the state-space models based on neural networks are presented in terms of the following discrete-time form [46]:

$$\boldsymbol{x}(k+1) = -\alpha\boldsymbol{x}(k) + \boldsymbol{f}\left(\boldsymbol{x}(k),\boldsymbol{u}(k)\right), \tag{2.34}$$

$$\boldsymbol{y}(k) = \boldsymbol{C}\boldsymbol{x}(k), \tag{2.35}$$

where $\alpha \in [-1, 1]$ is a fixed constant for controlling the state decaying. It should be noted that for $\alpha = 0$ the model (2.34) is equivalent to (2.29). Due to a different realization of the nonlinear function $\boldsymbol{f}(\cdot,\cdot)$ the following models can be distinguished [46]:

- the modified Hopfield type:

$$\boldsymbol{f}(\boldsymbol{x}(k),\boldsymbol{u}(k)) = \boldsymbol{A}\sigma(\boldsymbol{x}(k)) + \boldsymbol{B}\boldsymbol{u}(k), \tag{2.36}$$

- the modified Pineda type I:

$$\boldsymbol{f}(\boldsymbol{x}(k),\boldsymbol{u}(k)) = \sigma(\boldsymbol{A}\boldsymbol{x}(k) + \boldsymbol{B}\boldsymbol{u}(k)), \tag{2.37}$$

- the modified Pineda type II:

$$f(x(k), u(k)) = \sigma(Ax(k)) + Bu(k), \tag{2.38}$$

where A and B are weight matrices.

State-space models possess a number of advantages when compared to both fully and partially recurrent networks [34, 67]:

- The number of states (model order) can be selected independently of the number of hidden neurons. This way, only those neurons that feed their outputs back to the input layer through delays are responsible for defining the state of the network. As a result, the output neurons are excluded from the definition of the state.
- Since model states feed the input of the network, these are easily accessible from the external environment. This property can be useful when state measurements are available at some time instants (e.g. initial conditions).

In spite of the fact that state-space neural networks seem to be more promising than either fully or partially recurrent neural networks, in practice there is still a number of difficulties that can be encountered [67], e.g.

- the model states might not approach true process states,
- wrong initial conditions can deteriorate the performance, especially when short data sets are used for training,
- training can become unstable,
- the model itself after the training phrase can be unstable.

In particular, these drawbacks appear in cases when no state measurements and no initial conditions are available.

In brief, a very important property of the state-space neural network is that it can approximate a wide class of nonlinear dynamic systems [113]. There are, however, some restrictions. The approximation is only valid on compact subsets of the state-space and for finite time intervals, thus certain interesting dynamic characteristics are not reflected [34, 96].

Echo-state networks. The idea behind an echo-state network (ESN) is to make use of the advantages of a recurrent neural networks and at the same time to avoid problems of training a network when using traditional algorithms, such as back-propagation through time or the vanishing gradient. The ESN models provide an architecture and a supervised learning principle for recurrent neural networks (RNNs) [44, 61]. The main idea is as follows: (i) to drive a random, large, fixed recurrent neural network with the input signal, thereby inducing in each neuron within this "reservoir" network a nonlinear response signal, and (ii) to create a desired output signal by means of a trainable linear combination of all of the response signals. The structure of the ESN model is shown in Fig. 2.20.

The representation of the ESN model is provided by [44]:

$$x(k+1) = \sigma_h\left(W^x x(k) + W^u u(k+1) + W^y y(k)\right), \tag{2.39}$$

$$y(k) = \sigma_o\left(W^{out} z(k)\right), \tag{2.40}$$

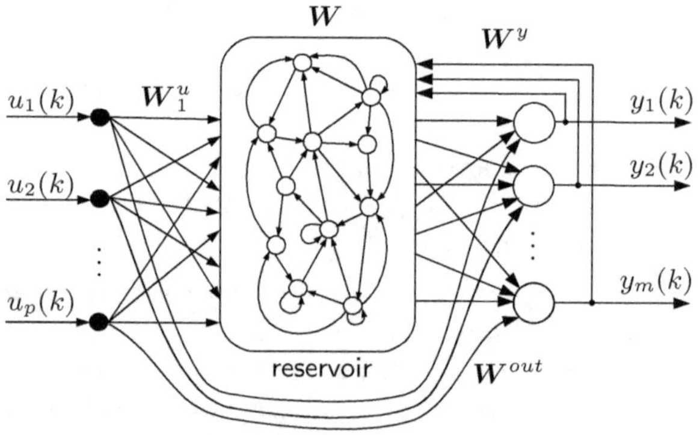

Fig. 2.20 A structure of an echo-state neural model

where $\boldsymbol{W}^x \in \mathbb{R}^{n\times n}$ is the reservoir weight matrix, $\boldsymbol{W}^u \in \mathbb{R}^{n\times p}$ is the input weight matrix, $\boldsymbol{W}^y \in \mathbb{R}^{n\times m}$ is the output feedback weight matrix, $\boldsymbol{W}^{out} \in \mathbb{R}^{m\times(n+p)}$ is the output weight matrix, $z(k) = [\boldsymbol{x}(k)\,\boldsymbol{u}(k)]^{\mathsf{T}}$ is augmented state vector, $\boldsymbol{\sigma}_h : \mathbb{R}^v \rightarrow \mathbb{R}^v$ and $\boldsymbol{\sigma}_o : \mathbb{R}^m \rightarrow \mathbb{R}^m$ are vector-valued activation functions of the hidden and output neurons, respectively (typically sigmoid or linear ones).

In order to obtain a sufficient approximation, the reservoir should be appropriately selected to provide a rich set of dynamic relations. A simple method to meet this demand is to use a reservoir that is sparsely and randomly connected. Sparse connections (only a few percent of possible connections) give a relative decoupling of sub-networks, which develop their individual dynamics. Typically, the reservoir consists of hundreds of neurons.

In order to ensure a proper functioning of ESN, the reservoir has to satisfy the echo state property that concerns the asymptotic properties of the reservoir dynamics excited by the driving signal. According to Proposition 3 of [44] the echo state property is guaranteed if the following condition is satisfied:

$$\sigma_{\max}(\boldsymbol{W}^x) < 1, \tag{2.41}$$

where $\sigma_{\max}$ is the largest singular value of the reservoir matrix $\boldsymbol{W}^x$.

The training process is significantly less complex compared to the case of the traditional recurrent neural networks, since it is only the output weight matrix $\boldsymbol{W}^{out}$ that is the subject of training. As discussed earlier in this section, the reservoir matrix $\boldsymbol{W}^x$ is generated randomly so that it satisfies (2.41). The input weight matrix $\boldsymbol{W}^u$ can be freely selected without violating the echo state property [44]. Most often, the input weights are chosen randomly. The same rules apply to the selection of the feedback weight matrix $\boldsymbol{W}^y$. The output weights are the linear regression weights of the desired outputs $\boldsymbol{y}_d(k)$ on the harvested extended states $z(k)$. Given the above, the simple off-line learning rule can now be formulated as:

$$\boldsymbol{W}^{out} = (\boldsymbol{X}^{-}\boldsymbol{Y})^{\mathsf{T}}, \tag{2.42}$$

where $\boldsymbol{X}$ is the state collection matrix obtained by feeding ESN with input sequence $\boldsymbol{u}(1), \ldots, \boldsymbol{u}(n)$, which yields a sequence $z(1), \ldots, z(n)$ of extended system states,

$\boldsymbol{Y}$ is the teacher output collection matrix composed of the desired outputs $\boldsymbol{y}_d(1), \ldots, \boldsymbol{y}_d(n)$, the operator $^-$ stands for the matrix pseudo-inverse. In recent years, the ESN models have been the subject of intensive research. The interested reader is referred to [3, 8, 10].

2.3.6 Spiking Neural Networks

Spiking neural networks belong to the third generation of the neural networks models. This class employs the concept of spiking neurons as the fundamental processing units [26, 60]. A spiking neuron is a more realistic neuron model compared to those used in the framework of the previous two generations of neural networks as discussed earlier in this chapter. As a result of using time as a resource for computation and communication, the spiking neurons much closer correspond to what the output of a biological neuron is. The best-known example of a spiking neuron is a leaky integrate-and-fire (IF) model represented by the following differential equation:

$$\tau_m \frac{dv(t)}{dt} = -v(t) + RI(t), \tag{2.43}$$

where $v(t)$ is the membrane potential, τ_m stands for the membrane time constant, R is the membrane resistance parameter, $I(t)$ represents the synaptic current. Spikes are formal events characterized by a firing time $t^{(f)}$. The firing time is defined by the threshold criterion:

$$t^{(f)} : v(t^{(f)}) = v_t, \tag{2.44}$$

where v_t is a threshold value. Immediately after $t^{(t)}$ the neuron potential is reset to a new value $v_r < v$ according to:

$$\lim_{t \to t^{(f)}} v(t) = v_r. \tag{2.45}$$

For $t > t^{(f)}$ the neuron dynamics is represented by (2.43). The combination of leaky integration (2.43) with reset (2.45) gives integrate-and-fire model. To date, many other spiking neurons have been developed, e.g. Hodgkin–Huxley, FitzHugh–Nagumo, Morris–Lecar, Izhikevich models, etc. For an exhaustive review of the spiking neuron models the interested reader is referred to [26, 43].

In order to implement the model (2.43), Euler's method of discretization can be easily employed. As a result, the following discrete-time IF model is obtained:

$$v(k+1) = \frac{\tau_m - \tau}{\tau} v(k) + \frac{\tau}{\tau_m} RI(k), \tag{2.46}$$

where τ is the integration time.

Let us now consider a simple case of a large and homogeneous population of N neurons. All neurons are identical and receive the same external input $I^{ext}(t)$. An interaction between ith and jth neuron is represented by the weight w_{ij} defined as:

$$w_{ij} = \frac{A}{N}, \tag{2.47}$$

where A is a parameter. The synaptic current of the ith neuron is represented by the formula:

$$I_i(t) = \sum_{j=1}^{N} \sum_{f} w_{ij} \alpha(t - t_j^{(f)}) + I^{ext}(t), \tag{2.48}$$

where $\alpha(t - t_j^{(f)})$ is the post-synaptic potential represented for a nonlinear function and the model (2.43) takes the form:

$$\tau_m \frac{dv_i(t)}{dt} = -v_i(t) + RI_i(t). \tag{2.49}$$

The connected large spiking neurons form what is known as the spiking neural network (SNN). The synaptic weights between neurons are then given by the matrix $W = \{w_{ij}\}$. The firing of the jth neuron changes the membrane potential of all the neurons by w_{ij} $(i = 1, \ldots, N)$.

In the last decade, SNNs have been successfully employed in control applications, especially in the context of robot control [9, 104, 109].

Pulse-coupled neural networks. A pulse-coupled neural network (PCNN) was proposed by Johnson and colleagues [47, 49] in 1993. The PCNN structure utilizes the Eckhorn's model of a neuron developed in 1989 to emulate the visual cortex of a cat [19]. PCNN is a single layer, two-dimensional, laterally-connected network of integrate-and-fire type neurons, with a one-to-one correspondence between the image pixels and network neurons. For this kind of a neural network no training is needed. Each pixel of an image is associated with a neuron of the PCNN model. The Eckhorn neuron in the framework of PCNN is illustrated in Fig. 2.21. In the neuron,

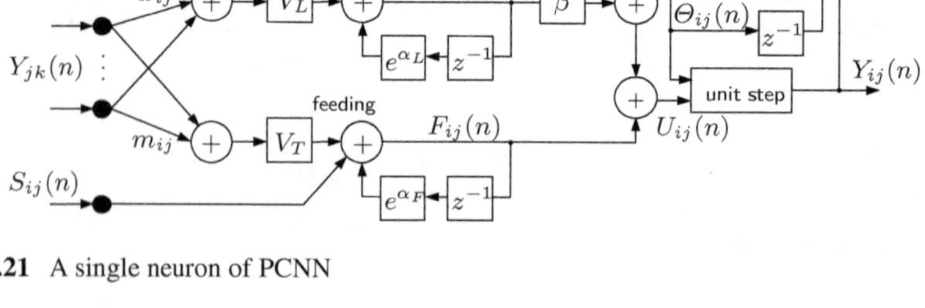

Fig. 2.21 A single neuron of PCNN

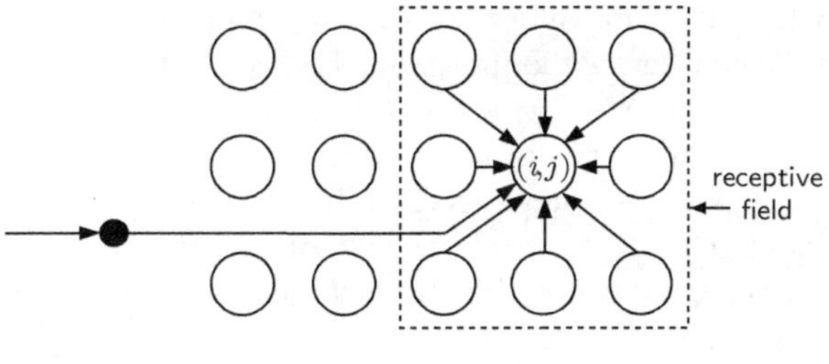

Fig. 2.22 An illustration of a neuron receptive field

one can distinguish three main parts: the input part, the linking module and the pulse generator. The neuron receives input signals from the feeding and the linking inputs. The feeding input is the primary input coming from the receptive area of the neuron. The receptive area (depicted in Fig. 2.22) simply includes the neighbouring pixels of the pixel (i, j) associated with the neuron. In turn, the linking input is the secondary input of lateral connections with the neighbouring neurons. The neuron is described by a set of equations. The linking module is represented as:

$$L_{ij}(n) = e^{-\alpha_L} L_{ij}(n-1) + V_L \sum_{kl} w_{ijkl} Y_{kl}(n-1), \tag{2.50}$$

where $L_{ij}(n)$ is the linking item of the neuron in (i, j) position at the time n, $Y_{kl}(n)$ stands for the pulse output of the neuron in (k, l) position, α_L is the attenuation time constant of the linking module, w_{ijkl} is the constant synaptic weight for the linking input, V_L stands for the inherent voltage potential of $L_{ij}(n)$. The feeding module is described by the formula:

$$F_{ij}(n) = e^{-\alpha_F} F_{ij}(n-1) + S_{ij} + V_F \sum_{kl} m_{ijkl} Y_{kl}(n-1), \tag{2.51}$$

where $F_{ij}(n)$ is the feedback input, α_F represents the attenuation time constant of the feeding module, m_{ijkl} is the constant synaptic weight for the feedback input, V_F is the inherent voltage potential of $F_{ij}(n)$ and S_{ij} is the input stimulus. The neuron activity is:

$$U_{ij}(n) = F_{ij}(n)\left(1 + \beta L_{ij}(n)\right), \tag{2.52}$$

where β is the linking coefficient. Finally, the output of the neuron can be written as follows:

$$Y_{ij}(n) = \begin{cases} 1 & \text{for } U_{ij}(n) > \Theta_{ij}(n) \\ 0 & \text{otherwise} \end{cases}, \tag{2.53}$$

where the threshold is defined as:

$$\Theta_{ij}(n) = e^{-\alpha_\Theta} \Theta_{ij}(n-1) + V_\Theta Y_{ij}(n), \tag{2.54}$$

where V_Θ is the inherent voltage potential of the threshold and α_Θ is the attenuation time constant of the threshold. The inter-connections between neurons, represented by matrices $M = \{m_{ijkl}\}$ and $W = \{w_{ijkl}\}$, are the constant synaptic weights that are dependent on the distance between neurons.

PCNN has been successfully applied in many academic and industrial fields including image processing (filtering, segmentation, coding) [58], biometric recognition, combinatorial optimization [59], feature extraction [59], encoding time sequences [48], etc.

2.3.7 A Long Short-Term Memory

A long short-term memory (LSTM) is a recurrent neural network proposed in 1997 by Hochreiter and Schmidhuber [38] as a solution for the problem of the blowing-up or vanishing gradient that was frequently observed when training the traditional recurrent models. The name long short-term relates to the fact that LSTM is a model for the short-term memory that can last for a long time. A common structure of LSTM consists of a memory cell, an input gate, an output gate and a forget gate. The memory cell is responsible for storing data for arbitrary time periods. In turn, gates control data flow through connections of the whole model. The basic representation of the single LSTM unit is given by the set of equations. The input gate is represented as:

$$\boldsymbol{i}(k) = \sigma_s(\boldsymbol{W}_i\boldsymbol{x}(k) + \boldsymbol{V}_i\boldsymbol{h}(k-1) + \boldsymbol{b}_i), \tag{2.55}$$

where $\boldsymbol{W}_i$, $\boldsymbol{V}_i$ are the weight matrices and $\boldsymbol{b}_i$ is the bias vector of the input gate, $\boldsymbol{x}(k)$ and $\boldsymbol{h}(k)$ are the input and the output of the LSTM unit, and σ_s is the activation function. In a similar manner, the output gate

$$\boldsymbol{o}(k) = \sigma_s(\boldsymbol{W}_o\boldsymbol{x}(k) + \boldsymbol{V}_o\boldsymbol{h}(k-1) + \boldsymbol{b}_o) \tag{2.56}$$

and the forget gate

$$\boldsymbol{f}(k) = \sigma_s(\boldsymbol{W}_f\boldsymbol{x}(k) + \boldsymbol{V}_f\boldsymbol{h}(k-1) + \boldsymbol{b}_f) \tag{2.57}$$

are described, where $\boldsymbol{W}_o$, $\boldsymbol{V}_o$ and $\boldsymbol{b}_o$ are parameters of the output gate and $\boldsymbol{W}_f$, $\boldsymbol{V}_f$ and $\boldsymbol{b}_f$ are parameters of the forget gate. Using (2.55)–(2.57), the memory cell is given by:

$$\boldsymbol{c}(k) = \boldsymbol{f} \circ \boldsymbol{c}(k-1) + \boldsymbol{i}(k) \circ \sigma_c(\boldsymbol{W}_c\boldsymbol{x}(k) + \boldsymbol{V}_c\boldsymbol{h}(k-1) + \boldsymbol{b}_c) \tag{2.58}$$

and the cell output:

$$\boldsymbol{h}(k) = \boldsymbol{o}(k) \circ \sigma_h(\boldsymbol{c}(k)), \tag{2.59}$$

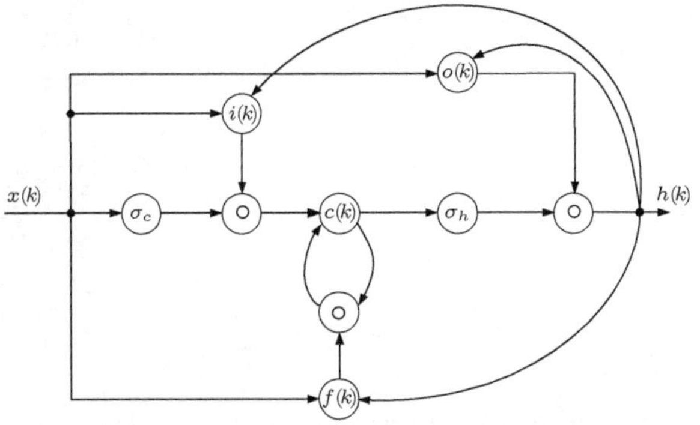

Fig. 2.23 A simplified structure of an LSTM unit (∘ – Hadamard product)

where $\boldsymbol{W}_c$, $\boldsymbol{V}_c$ and $\boldsymbol{b}_c$ are weight matrices and bias vector of the memory cell, σ_c and σ_h stand for the activation functions, ∘ is the Hadamard product (element-wise product), and the initial values are $\boldsymbol{c}(0) = \boldsymbol{0}$ and $\boldsymbol{h}(0) = \boldsymbol{0}$.

Either a memory cell or gate units can convey useful information about the current state of the network. The memory cell stores the network state, for either a long or a short period. This is achieved by means of the identity activation function. The input gate controls the extent, to which a new value (or a state) flows into the memory cell. The forget gate controls the extent, to which a value remains in the memory cell; and finally the output gate controls the extent, to which the value in the memory cell is used to compute the output of the memory cell. A single LSTM unit is shown in Fig. 2.23. One can design larger networks using the LSTM units. The topology of the entire network consists of one input layer, one hidden layer and one output layer. The fully self-connected hidden layer includes a number of memory cells and corresponding gate units. For structural details the reader is referred to the work [38]. Such a neural network can be trained using standard algorithms dedicated for recurrent neural networks, e.g. RTRL or truncated BPTT mentioned in Sect. 2.3.2.

2.4 Developing Models

2.4.1 Forward Modelling

Let us assume that a plant is governed by the following nonlinear discrete difference equation:

$$y(k+1) = f(y(k), \ldots, y(k-n_a+1), u(k), \ldots, u(k-n_b+1)). \tag{2.60}$$

The process output depends on its n_a past output and n_b inputs, in the sense defined by the nonlinear input-output relation $f(\cdot)$.

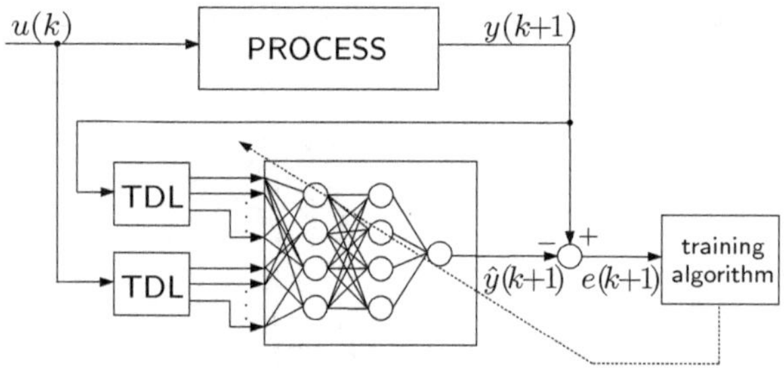

Fig. 2.24 A series-parallel identification scheme

Series-parallel identification. In order to train the neural networks with external dynamics of the NARX or NAR type, the real plant outputs can be used. The corresponding model is known as a series-parallel identification model (depicted in Fig. 2.24) and is described by the following difference equation [66, 71]:

$$\hat{y}(k+1) = \hat{f}(y(k), \ldots, y(k-n_a+1), u(k), \ldots, u(k-n_b+1)). \tag{2.61}$$

The time delay lines, visible in Fig. 2.24, provide all the signals required for the designing of a model. On the basis of the output error $e(k)$, a training algorithm should adjust the network parameters to minimize the performance index determined using the output error. In a general case, the identification problem can be defined as finding a process output-input mapping based on the input-output data pairs. The process is treated like a "black box", assuming that the input and output signals are known. Such identification scheme is often referred to as a non-parametric. The model should be trained to mimic the process behaviour as close as possible.

The series-parallel identification model (2.61) guarantees the stability of the training process and makes it possible to apply the standard training algorithms developed for the feed-forward networks. In (2.61) $\hat{f}$ represents a nonlinear input-output relation captured by the neural network (in the sense of approximation of the relation $f(\cdot)$). Note that the input of the neural network in this case contains the past values of process output. The neural network structure is quite simple as it has no internal feedbacks, however the network input space is relatively large since all the necessary past inputs and outputs are to be feed to the network.

Assuming that the neural model (2.61) has been trained up to a given accuracy and faithfully represents a given process ($\hat{y}(k) \approx y(k)$), the neural model can be described by (2.62) [41]. In this way, the suitably adjusted neural network can be used independently for modelling. In fact, the neural network described by the formula (2.62) uses its own output values as part of the input space. Thus, the feedback from the network output to its input is introduced and the feed-forward network becomes a recurrent network with outer feedback.

Parallel identification. The series-parallel identification model cannot be used to train the NOE model, since it requires (its own) previous outputs, thus establishing the output feedbacks. In this case, the following difference model is obtained:

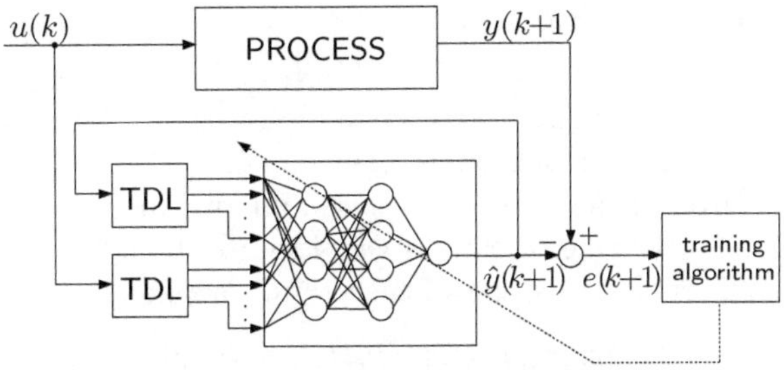

Fig. 2.25 A parallel identification scheme

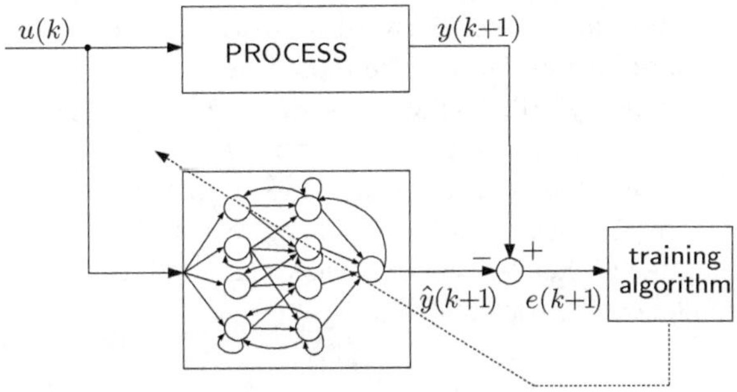

Fig. 2.26 A parallel identification scheme for recurrent networks

$$\hat{y}(k+1) = \hat{f}(\hat{y}(k), \ldots, \hat{y}(k-n_a+1), u(k), \ldots, u(k-n_b+1)), \qquad (2.62)$$

which is trained using the parallel identification model [66, 71]. Unfortunately, owing to feedbacks, a training algorithm should also take into account the dynamic relations among the network variables. This difficulty leads to relatively complicated training procedures. Moreover, taking into account the limitations concerning the input and output signals, the stability of the considered parallel structure cannot be guaranteed. Therefore, one might be faced with the situation where the training process does not converge and the output error will not converge to zero. The general idea behind the parallel identification scheme is shown in Fig. 2.25. This scheme can be easily extended to train the NARMAX models. The only thing necessary to its implementation is to feed the prediction error back to the network input according to the description provided in Sect. 2.3.1. All the previous values of the prediction error needed are provided by the TDL of length n_c.

The parallel identification scheme can be also employed to design models based on recurrent networks. A common feature of all the considered recurrent neural networks is that during the training phase such networks only require input and output data at a current time step. The past input and output data are restored within the network structure. Thus, the training can be carried out using the scheme depicted in Fig. 2.26. What is important, the parallel identification scheme shown in Fig. 2.26 can be used not only to identify input-output neural models but also in the context of the state-space models described in Sect. 2.3.5.

2.4.2 Inverse Modelling

The inverse models of dynamic systems play a fundamental role in control structures. The inverse model can be applied in the well-known control schemes: direct inverse control, model reference control, internal model control or predictive control [41]. Similarly to the case of forward identification, the models with external dynamics or recurrent networks can be successfully applied in the context of inverse identification.

General learning architecture. The training idea depicted in Fig 2.27 is known as the general learning architecture [68, 86]. The process output $y(k+1)$ is used as the input to the feed-forward network. The network output $\hat{u}(k)$ is compared with the process input $u(k)$. The input error is used to train the network. Let us consider the model (2.60). It is obvious, that the inverse function $\hat{f}^{-1}$ allows one to generate $\hat{u}(k)$, which depends on the future value of $y(k+1)$. To overcome this problem, this signal can be replaced with the reference value $r(k+1)$, assuming that $r(k+1)$ is available at the time k. This is a reasonable assumption, since a reference signal is typically known a priori. Thus, the nonlinear input-output relation of the inverse model can be formulated as:

$$\hat{u}(k) = \hat{f}^{-1}(r(k+1), y(k), \ldots, y(k-n_a+1), u(k-1), \ldots, u(k-n_b+1)). \quad (2.63)$$

The current and past system outputs, the training (reference) signal, and the past values of the system input are the inputs to the inverse neural network model. The neural network is trained off-line in order to minimize a criterion based on the difference between the process input $u(k)$ and estimated input $\hat{u}(k)$, e.g.

$$J = \sum_{k=1}^{N} (u(k) - \hat{u}(k))^2, \quad (2.64)$$

where N is the length of the reference signal. The success of these learning schemes is related to the ability of the neural network to generalize the knowledge. This

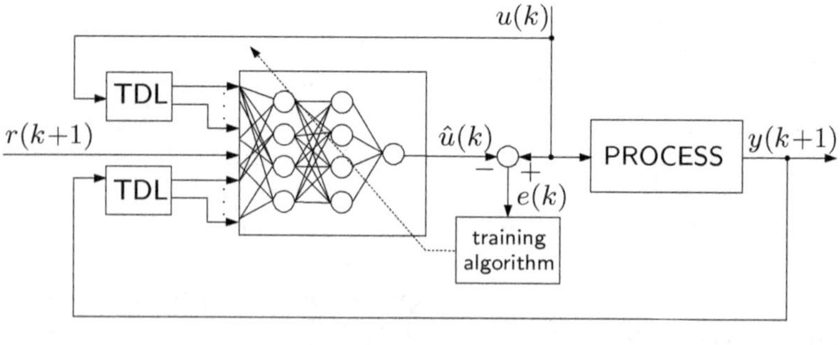

Fig. 2.27 Inverse modelling using external dynamic neural networks. Generalized training

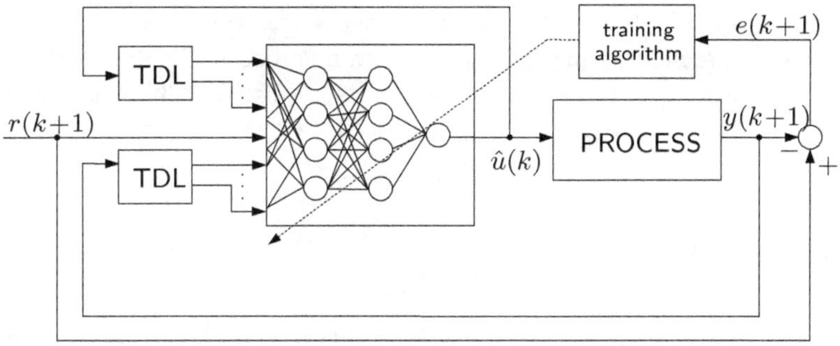

Fig. 2.28 Inverse modelling using external dynamic neural networks. Specialized training

architecture cannot selectively train the model to respond correctly in regions of interest because it is unknown which plant inputs $u(k)$ correspond to the desired outputs $y(k+1)$ [86]. Thus, the problem with the general learning is that the network should be trained using the realistic data. However, the realistic data can be recorded only when the initially trained inverse model is used as a controller. It is desirable to repeat the model design in the loop and the inverse model can be retrained with the newly recorded data. This problem is referred to as *identification for control* [68].

Specialized learning architecture. In order to deal with the *identification-for-control* problem, a specialized learning architecture was developed in [86]. A block scheme of this kind of inverse modelling is shown in Fig. 2.28. The scheme makes it possible for a proper operation of the neural controller in regions of specialization. The neural network is trained to find the process input $\hat{u}(k)$ that drives the process output $y(k+1)$ to the desired value $r(k+1)$. In this way, the controller can be designed to achieve an acceptable control performance, e.g. near zero steady-state error. Here, the neural network is trained on-line in order to minimize the criterion based on the difference between the process output $y(k+1)$ and the reference $r(k+1)$, e.g.

$$J = \sum_{k=1}^{N} (r(k) - y(k))^2. \tag{2.65}$$

In order to minimize (2.65), some well-known training algorithms (e.g. gradient descent) can be employed. However, gradient-based learning procedures require knowledge of the Jacobian of the process. The Jacobian of the process is usually unknown as a mathematical model of the process is unknown. A possible solution is to identify the process to provide the estimate of the Jacobian. In this context, the forward models based on neural networks can be successfully used [68].

It is in fact possible to combine the general and specialized learning. In the first stage, a general training is carried out off-line to learn the approximate behaviour of the process. This gives the initial weights for the specialized training. In the second stage, a specialized training is performed on-line to fine-tune the network in the operating regime of the process. The second stage is realized in a closed-loop control. Such a procedure can speed up the learning process by reducing the number of iterations during the specialized training phase. Moreover, the neural controller can adapt more easily, if the operating points of the process do change.

2.5 Robust Models

In the recent years, a great emphasis has been put on providing uncertainty descriptions for the models used for control purposes. These problems can be referred to as robust identification. The robust identification procedure should deliver not only a model of a given process but also a reliable estimate of uncertainty associated with the model. There are three main factors that contribute to uncertainty in models fitted to data [87]:

- noise corrupting the data,
- changing plant dynamics,
- selecting a model that cannot capture the true process dynamics.

The two main philosophies for addressing the above found in the literature are as follows:

1. *Bounded error approaches* or *set-membership identification* [64, 103]. This group of approaches relies on the assumption that the identification error is unknown but bounded. In such a case, the identification provides hard error bounds, which guarantee upper bounds on model uncertainty [29]. In this framework, the property of robustness is closely related to the characteristics of the model identification process;
2. *Statistical error bounds*. In the case of such approaches, statistical methods are used to quantify model uncertainty by means of the so-called soft error bounds. In this framework, the identification is carried out without taking into account the question of robustness, which is dealt with in a separate step. This usually leads to the least-squares estimation and prediction error methods [88].

2.5.1 *Nonlinear Set-Membership Identification*

In order to identify the uncertainty associated with the model, the set-membership (SM) method (used in linear systems identification) can be employed. The important characteristic of this approach is that no assumptions on the functional form of the system is required. The method assumes the noise to be unknown but bounded. Let us consider the system as represented by:

$$\tilde{y}(k+1) = f(\tilde{\varphi}(k)) + d(k), \tag{2.66}$$

where $f(\cdot)$ is an unknown nonlinear function $\tilde{y}$ and $\tilde{\varphi}(k)$ are the (noise-corrupted) output and the regression vector, respectively with $d(k)$ standing for the noise satisfying

$$|d(k)| \leq \varepsilon(k), \ \forall k, \tag{2.67}$$

where $\varepsilon(k)$ is the upper error bound. It is also assumed that $f(\cdot)$ satisfies the conditions:

$$f(\varphi(k)) \in \mathcal{F}(\gamma) \doteq \{f(\varphi(k)) \in C^1 : \|f'(\varphi(k))\| \leq \gamma \ \ \forall\, \varphi(k) \in \Phi\}, \tag{2.68}$$

where $f'(\varphi(k))$ is the gradient of $f(\varphi(k))$ and Ψ is the set of possible regressors. The SM approach described here takes into account the rate of variation of the function $f(\cdot)$. As we deal with nonlinear systems, a function $f(\cdot)$ can be realized using the neural networks of the dynamic kind described earlier in this chapter.

The key role in the set-membership identification is played by the *feasible system set* (FSS) (or *unfalsified system set*). FSS is defined as follows:

$$FSS \doteq \{\hat{f}(\varphi(k)) \in \mathcal{F}(\gamma) : |\tilde{y}(k+1) - \hat{f}(\tilde{\varphi}(k))| \leq \varepsilon(k),\ k = 1, \ldots, N\}. \tag{2.69}$$

A feasible system set summarizes all the information on data generation available up to time N. If the considered system $f \in FFS$ then

$$\underline{f}(\varphi(k)) \leq f(\varphi(k)) \leq \overline{f}(\varphi(k)), \quad \forall \varphi(k) \in \Phi, \tag{2.70}$$

where

$$\overline{f}(\varphi(k)) = \sup_{f \in FSS} f(\varphi(k)) \ \text{ and } \ \underline{f}(\varphi(k)) = \inf_{f \in FSS} f(\varphi(k)). \tag{2.71}$$

If all assumptions hold, $\overline{f}(\varphi(k))$ and $\underline{f}(\varphi(k))$ are the tightest upper and lower bounds of $f(\varphi(k))$ and are called the optimal bounds [64]. A necessary condition to guarantee that $FSS \neq \varnothing$ is:

$$f_u(\tilde{\varphi}(k)) \geq \tilde{y}(k+1) - \varepsilon(k), \quad k = 1 \ldots, N, \tag{2.72}$$

where

$$f_u(\tilde{\varphi}(k)) \doteq \min_{k=1,\ldots,N} \left(\tilde{y}(k+1) + \varepsilon(k) + \gamma\|\varphi(k) - \tilde{\varphi}(k)\|\right). \tag{2.73}$$

A sufficient condition for obtaining $FSS \neq \varnothing$ is

$$f_u(\tilde{\varphi}(k)) > \tilde{y}(k+1) - \varepsilon(k), \quad k = 1 \ldots, N. \tag{2.74}$$

As pointed out in [64], there is no gap between the necessary and sufficient conditions, since a condition

$$f_u(\tilde{\varphi}(k)) > \tilde{y}(k+1) - \varepsilon(k) + \delta, \quad k = 1\ldots, N \tag{2.75}$$

is sufficient for any $\delta > 0$ arbitrarily small and necessary for $\delta = 0$. In order to satisfy the sufficient condition (2.74), both γ and $\varepsilon(k)$ should be suitably modified. Therefore, let us introduce the relative plus absolute model of the error bound in the form:

$$|d(k)| \leq \varepsilon(k) = \varepsilon_r |y(k+1)| + \varepsilon_a, \quad \varepsilon_r, \varepsilon_a \geq 0. \tag{2.76}$$

The function

$$\gamma^*(\varepsilon_r, \varepsilon_a) \doteq \inf_{\substack{\varepsilon_r, \varepsilon_a, \gamma \\ FSS \neq \varnothing}} \gamma \tag{2.77}$$

describes a surface that separates the falsified values of $\varepsilon_r, \varepsilon_a$ and γ from the validated ones. All the parameters $\varepsilon_r, \varepsilon_a$ and γ need to be selected in the validated region. There are some useful rules for selecting γ and ε_a [64]. The parameter γ can be selected taking into account the gradient of $f(\varphi(k))$. In turn, ε_a can simply be chosen as the accuracy of a sensor used for measurements acquisition. Analogously to (2.73) the lower bound of the function $f(\varphi(k))$ can be defined:

$$f_l(\tilde{\varphi}(k)) \doteq \max_{k=1,\ldots,N} \left(\tilde{y}(k+1) - \varepsilon(k) - \gamma\|\varphi(k) - \tilde{\varphi}(k)\|\right). \tag{2.78}$$

The functions (2.73) and (2.78) are optimal bounds, i.e.

$$\overline{f}(\varphi(k)) \doteq f_u(\tilde{\varphi}(k)), \quad \underline{f}(\varphi(k)) \doteq f_l(\tilde{\varphi}(k)). \tag{2.79}$$

It means that the smallest interval guaranteed to include $f(\varphi(k))$ is the interval $[\underline{f}(\varphi(k)), \overline{f}(\varphi(k))]$. Finally, the robust estimate of $f(\varphi(k))$ is given by:

$$f_c(\varphi(k)) = \frac{1}{2}[\underline{f}(\varphi(k)) + \overline{f}(\varphi(k))]. \tag{2.80}$$

Millanese and Novarra also investigated the problem of selecting the suitable scaling of the regressors to better adapt the model to data. They proposed to estimate quantities:

$$\mu_i = \max_{\varphi(k)\in\Phi} \left|\frac{\partial f(\varphi(k))}{\partial \varphi_i}\right|, \quad i = 1, \ldots, n, \tag{2.81}$$

where φ_i denotes the ith component of the vector $\varphi(k)$. Then the optimal outer approximation of the function $f(\cdot)$ can be realized by factors:

$$\nu_i = \frac{1}{n\mu_i^2}, \quad i = 1, \ldots, n, \tag{2.82}$$

used to regressors scaling. The whole procedure is summarized by Algorithm 2.1, where a global bound on the derivative of $f(\varphi(k))$ was assumed. However, to obtain a more accurate identification a different bounds γ_k on suitable partition of Ψ could be considered. Local assumptions on $f(\varphi(k))$ can be achieved by application of the residue function:

$$f_{\Delta}(\varphi(k)) = f(\varphi(k)) - \hat{f}(\varphi(k)) \quad (2.83)$$

by the set of data

$$\Delta y(k+1) = \tilde{y}(k+1) + f_{\Delta}^{c}(\varphi(k)), \quad (2.84)$$

where $f_{\Delta}^{c}(\varphi(k))$ is the central estimate of $f_{\Delta}(\varphi(k))$ obtained using data $\Delta y(k+1)$. Local identification method makes it possible to derive the following uncertainty bounds:

$$\hat{y}(\varphi(k)) + \underline{f}_{\Delta}(\varphi(k)) \leq f(\varphi(k)) \leq \hat{y}(\varphi(k)) + \overline{f}_{\Delta}(\varphi(k)), \forall \varphi(k) \in \Phi. \quad (2.85)$$

Similarly to the case of the global method, regressor scaling can be employed to improve the identification accuracy. The local set-membership identification is presented by Algorithm 2.2.

2.5.2 Model Error Modelling

Prediction-error approaches are widely used in designing empirical process models for control purposes and fault diagnosis. Great emphasis has been put on providing uncertainty descriptions. In control theory, the identification that provides the uncertainty of the model is referred to as the control-relevant identification or robust identification [18, 87, 88]. In order to characterize uncertainty in the model, an estimate of the target model is required. To obtain the latter, a model of an increasing complexity is designed up to the point, where it is falsified (the hypothesis that the model provides an adequate description of a process is accepted at a selected significance level). Certain statistical tools are then used to derive uncertainty in the parameters. The model-error modelling employs prediction error methods in order to identify a model from the input-output data [72, 88]. When this is done, one is able to estimate the uncertainty of the model by analyzing the residuals evaluated from the inputs. Uncertainty is a measure of unmodelled dynamics, noise and disturbances. The identification of residuals provides the so-called *model-error model*. In the original algorithm, a nominal model along with uncertainty is constructed by adding the outputs of the nominal and error models for each frequency value. As discussed in [88], the key problem is to find a proper structure for the error model. One can simply start with an a priori chosen flexible structure, e.g. the tenth-order FIR filter. If this error model is not falsified by the data, it has to be kept. Otherwise, the model complexity should be increased until it is unfalsified by the data.

An algorithm allowing one to establish uncertainty bounds in the time domain was proposed in [70, 74]. It was intended to be used in the fault-diagnosis framework. The model-error modelling scheme can be carried out using neural networks of the

Algorithm 2.1 Nonlinear global set-membership identification

Step 1. Partition data into estimation and calibration data sets. Determine the range of each element of the regression vector $\varphi_i \in [\underline{\varphi}_i, \overline{\varphi}_i]$ $i = 1, \ldots, n$ giving

$$\Phi = \left\{\left[\underline{\varphi}_1, \overline{\varphi}_1\right] \times \cdots \times \left[\underline{\varphi}_n, \overline{\varphi}_n\right]\right\}.$$

Step 2. Train a neural network model $\hat{f}(\varphi(k))$ using the estimation data set.

Step 3. Calculate

$$\mu_i = \max_{\varphi \in \Phi} \left| \frac{\partial \hat{f}(\varphi(k))}{\partial \varphi_i} \right|, \quad i = 1, \ldots, n$$

then determine the scaled regressors

$$\phi_i = \frac{\varphi_i}{\sqrt{\nu_i}}, \quad \nu_i = \frac{1}{n\mu_i^2},$$

where

$$\Phi_s = \left\{\left[\frac{\underline{\varphi}_1}{\sqrt{\nu_1}}, \frac{\overline{\varphi}_1}{\sqrt{\nu_1}}\right] \times \cdots \times \left[\frac{\underline{\varphi}_n}{\sqrt{\nu_n}}, \frac{\overline{\varphi}_n}{\sqrt{\nu_n}}\right]\right\}.$$

Step 4. Compute the surface $\gamma^*(\varepsilon_r, \varepsilon_a)$ according to (2.77) on the suitable range of ε_r and ε_a. During this task the scaled regressors should be used.

Step 5. Choose γ, ε_r and ε_a, e.g. $\hat{\gamma} \cong \max_{\phi(k) \in \Phi_s} \|\hat{f}'(\phi(k))\|$, $\hat{\varepsilon}_a \cong$ is the accuracy of the measuring device, and ε_r should be selected in such a way as to obtain the minimum value of a simulation error calculated over the calibration data set using the regression model designed as in Step 6.

Step 6. Construct a robust model as

$$\hat{y}(k+1) = f_c(\phi(k)),$$

where

$$f_c(\phi(k)) = \frac{1}{2}\left(\underline{f}(\phi(k)) + \overline{f}(\phi(k))\right),$$

where $\underline{f}(\phi(k))$ and $\overline{f}(\phi(k))$ are represented by (2.78) and (2.73) respectively.

dynamic type, as discussed in Sect. 2.3. For the sake of simplicity, let us assume that both the fundamental model of the process and the error model can be modelled using neural networks of the NARX type. The first stage is to construct the fundamental model of the process:

$$\hat{y}(k+1) = \hat{f}(\varphi(k)), \tag{2.88}$$

where

$$\varphi(k) = [y(k), \ldots, y(k-n_a+1), u(k), \ldots, u(k-n_b+1)]^{\mathsf{T}}. \tag{2.89}$$

Algorithm 2.2 Nonlinear local set-membership identification

Step 1. Partition data into estimation and calibration data sets. Determine the range of each element of the regression vector.
Step 2. Train a neural network model $\hat{f}(\varphi(k))$ using the estimation data set.
Step 3. Consider the residue function

$$f_\Delta(\varphi(k)) = f(\varphi(k)) - \hat{f}(\varphi(k)) \tag{2.86}$$

and record the set of data

$$\Delta y(k+1) = \tilde{y}(k+1) - \hat{f}(\tilde{\varphi}(k)), \tag{2.87}$$

Step 4. Train a neural network model $\hat{f}_\Delta(\varphi(k))$.
Step 5. For the model $\hat{f}_\Delta(\varphi(k))$ determine the scaled regressors $\phi(k)$ analogously as in Step 3 of Algorithm 1.
Step 6. Compute the surface $\gamma^*(\varepsilon_r, \varepsilon_a)$ according to (2.77) on the suitable range of ε_r and ε_a for the model with the scaled regressors $\phi(k)$.
Step 7. Using calibration set choose γ, ε_r and ε_a, in the same way as in Step 5 of Algorithm 1.
Step 8. Construct a robust model as

$$\hat{y}(k+1) = \hat{f}(\varphi(k)) + f_\Delta^c(\phi(k)),$$

where the central estimate $f_\Delta^c(\phi(k))$ can be derived in the similar fashion as in global case.

The next step is to design an error model. In order to collect the data, the residual $r(k+1)$ is computed according to:

$$r(k+1) = y(k+1) - \hat{y}(k+1). \tag{2.90}$$

In this approach, a neural network is used to model an "error" system with the input $u(k)$ and the output $r(k+1)$ as follows:

$$y_e(k+1) = f_e(\phi(k)), \tag{2.91}$$

where

$$\phi(k) = [r(k), \ldots, r(k-n_c+1), u(k), \ldots, u(k-n_d+1)]^\mathsf{T} \tag{2.92}$$

and n_c and n_d are number of delayed residuals and inputs, respectively. After the training, the response of the error model is used to form the uncertainty bounds, where the centre of the uncertainty region is defined as a sum of the output of the system model and the output of the error model. Then, the upper band can be calculated as:

$$\overline{y}(k) = \hat{y}(k) + y_e(k) + t_\alpha \sigma, \tag{2.93}$$

and the lower band in the following way:

$$\underline{y}(k) = \hat{y}(k) + y_e(k) - t_\alpha \sigma, \tag{2.94}$$

where t_α is the $\mathcal{N}(0, 1)$ tabulated value assigned to the confidence level, e.g. $\alpha = 0.05$ or $\alpha = 0.01$, σ is the standard deviation of $y_e(k)$. It should be kept in mind that the error model represents not only the residual but also structured uncertainty, disturbances, etc. Therefore, the uncertainty bounds (2.93) and (2.94) should work well only under the assumption that the signal $y_e(k)$ has a normal distribution. The centre of the uncertainty region is the signal $\hat{y}(k) + y_e(k) \approx y(k)$. The designing procedure is described by Algorithm 2.3. Recently, MEM has been successfully applied to the robust model-predictive control [73] as well as to the fault-tolerant control [16].

2.5.3 Statistical Bounds

Let us assume that the model output contains some additive noise $w(k)$ (most often assumed to be the realization of the zero-mean, Gaussian and white stochastic process) [4, 23]:

$$y(k) = \hat{y}(k) + w(k), \tag{2.95}$$

where $y(k)$ is the measured system output. In such a scenario, all the measurement errors and disturbances can be represented by the component $w(k)$. Such a representation of uncertainties is often referred to as global uncertainties [11]. This is the general way to express uncertainties for nonlinear systems described by input-output models. In what follows, a procedure for estimating the uncertainty region using optimum experimental design approach is proposed and it is also shown how input data recorded in a plant can influence model parameters variations observed during model training. In the following, optimum experimental design (OED) is employed to derive uncertainty bounds of the neural network model as proposed in the work of Patan and colleagues [78].

Let $y^j = y(u^j; \theta) = \{y(k; \theta)\}_{k=0}^{L_j}$ denote the sequence of neural network responses for the sequence of inputs $u^j = \{u(k)\}_{k=0}^{L_j}$ related to the consecutive time moments $k = 0, \ldots, L_j < \infty$ and selected from among an a priori given set of input sequences $\mathcal{U} = \{u^1, \ldots, u^P\}$. Here θ represents a vector formed from all unknown network parameters that have to be estimated via training process using observations of the system. In order to select the most informative data sequences for the training of the neural network, a quantitative measure of the goodness of model parameter identification is required. A reasonable approach is to choose a performance measure defined on the Fisher information matrix (FIM) associated with the parameters to be estimated, the inverse of which constitutes the lower bound of the covariance matrix for the parameter estimates [77, 80]. Introducing for each possible sequence u^i

Algorithm 2.3 Model error modelling

Step 1. Partition data into training and testing data sets.

Step 2. Train a neural network model $\hat{f}(\varphi(k))$ using the training data set. If the NARX model is used for this purpose, the regression vector is

$$\varphi(k) = [y(k), \ldots, y(k - n_a + 1), u(k), \ldots, u(k - n_b + 1)]^{\mathsf{T}}.$$

Step 3. Test the model $\hat{f}(\varphi(k))$ on the testing data set and compute the residue function

$$r(k + 1) = y(k + 1) - \hat{f}(\varphi(k)),$$

Step 4. Collect the data $\{u(k), r(k)\}_1^N$ and identify an error model $f_e(\phi(k))$ using these data. If the NARX model is used to represent $f_e(\phi(k))$, the regression vector is

$$\phi(k) = [r(k), \ldots, r(k - n_d + 1), u(k), \ldots, u(k - n_c + 1)]^{\mathsf{T}}.$$

Step 5. Derive the centre of the uncertainty region as

$$y(k + 1) = \hat{f}(\varphi(k)) + f_e(\phi(k)).$$

Step 6. Determine uncertainty bounds according to (2.94) and (2.93).

$(i = 1, \ldots, P)$ a variable v_i taking the value 1 or 0 depending on whether a sequence is chosen or not. The FIM can be written as [77]:

$$M(v_1, \ldots, v_P) = \sum_{i=1}^{P} \frac{v_i}{SL_i} \sum_{k=0}^{L_i} \left(\frac{\partial y(u^i, k; \theta)}{\partial \theta}\right)\left(\frac{\partial y(u^i, k; \theta)}{\partial \theta}\right)^{\mathsf{T}}\Bigg|_{\theta=\theta^0}, \tag{2.96}$$

with θ^0 being a prior estimate to the θ that can be obtained from the previous experiments or alternatively some known nominal values can be used [81, 102]. As for the criterion, various choices are proposed in the literature but the most commonly used one is the so-called D-optimality (determinant) criterion, which minimizes the volume of the uncertainty ellipsoid for the parameter estimates [4, 80]. In such a case, our design problem consist in finding a sequence $v = (v_1, \ldots, v_P)$ that solves the problem:

$$\begin{aligned} \max \quad & \mathcal{P}(v) = \log\det\big(M(v)\big), \\ \text{s.t.} \quad & v_i = 0 \text{ or } 1, \quad i = 1, \ldots, P, \\ & \sum_{i=1}^{P} v_i = S. \end{aligned} \tag{2.97}$$

This constitutes a 0–1 integer programming problem. As for its approximate solution, a very efficient exchange algorithm can be easily adopted based on the notion of the so-called restricted-design measures (cf. [79, 80, 101] for details) originating from the experimental design theory. What is essential in describing the uncertainty of the model is to find the relation between the variability of the model parameters θ and the variability of the network output $\hat{y}(\cdot)$. It can be provided in the form of the following result:

Theorem 2.1 ([78]) *The sequence* $v^\star$ *maximizes* $\mathcal{P}(v) = \log\det M(v)$ *iff* $v^\star$ *minimizes* $\max_{u^i \in \mathcal{U}} \phi(u^i, v)$*, where*

$$\phi(u^i, v) = \operatorname{trace}\left(\frac{1}{L_i}\sum_{k=0}^{L_i}\left(\frac{\partial y(u^i, k; \theta)}{\partial \theta}\right)^T M^{-1}(v)\left(\frac{\partial y(u^i, k; \theta)}{\partial \theta}\right)\right). \tag{2.98}$$

Theorem 2.1 is also a consequence of more general considerations provided in [80, Thm. 3.1] or [81, Thm. 2]. The function $\phi(\cdot, \cdot)$ is of paramount importance here as it may be interpreted in terms of a normalized variance of the predicted network response and define the variability of the system output. Immediately we obtain that that through maximization of the D-optimality criterion, the maximal level of the prediction variance is suppressed. Thus, the optimal data sequences decrease the maximal variability of the model response, i.e. the uncertainty of the model is significantly reduced.

Once the network is trained with the optimal data and the variance of the predicted network response is determined, one is able to define the model uncertainty region. In particular, owing to the assumptions imposed on the uncertainty $w(k)$, we can make use of the interval estimation methods. For a given significance level α, which represents a fixed range of model uncertainty, the uncertainty has the form:

$$w(k) = k_\alpha \phi(u, k, v^\star), \tag{2.99}$$

where

$$\phi(u, k, v^\star) = \left(\frac{\partial y(u, k; \theta)}{\partial \theta}\right)^T M^{-1}(v^\star)\left(\frac{\partial y(u, k; \theta)}{\partial \theta}\right) \tag{2.100}$$

and k_α is such that $100(1-\alpha)\%$ of the distribution lies within the bounds. The meaning of this is that with the significance level α, all possible output values are included in the region defined by a lower

$$w_l(k) = \hat{y}(k) - w(k) \tag{2.101}$$

and an upper

$$w_u(k) = \hat{y}(k) + w(k) \tag{2.102}$$

uncertainty estimates. This can be further used as a decision rule for defining the level of model robustness. The proposed procedure is summarized by Algorithm 2.4.

Algorithm 2.4 Statistical bounds

Step 1. Partition data into P data sets of the same length. Set the final number of training sequences S. Set the confidence level α.

Step 2. Preliminary train a neural network model using any data set to obtain prior estimate of network parameters θ^0.

Step 3. Solve (2.97) and determine the most informative training data sets.

Step 4. Train the neural network model using the selected data sets.

Step 5. Calculate the neural model uncertainty according to:

$$w(k) = k_\alpha \phi(u, k, v^\star),$$

where k_α is a tabulated value assigned to α confidence level and $\phi(\cdot, \cdot, \cdot)$ is defined by (2.100).

Step 6. Determine uncertainty bounds according to (2.101) and (2.102).

2.6 Conclusions

This chapter described different neural network architectures that can be used in the context of control applications. Keeping the described classification of neural networks in mind, one sees that it is the second generation that is now commonly used for the purposes of control systems synthesis. We discussed a number of possible realizations discussed, starting from simple feed-forward networks and up to more sophisticated recurrent architectures. Obviously, each of these structures has some advantages and disadvantages. Feed-forward networks are attractive owing to their simplicity and good approximation abilities. On the other hand, the globally recurrent networks have a more complex structure but represent better natural dynamic behaviour. In turn, the self-organizing map is quite a different neural structure that can adapt its parameters using the input data (unsupervised learning). The third generation of the neural network models, covering such models as the so-called impulsive and the spiking ones, can be viewed as a new, and very promising, trend in neural network modelling. The chapter also provided a review of the learning schemes for obtaining a proper neural network model, e.g. the forward and the inverse modelling. Moreover, we also described learning carried out in both static and dynamic way. We can choose the appropriate learning scheme allowing one to obtain the most accurate model for the purposes of the specific control system synthesis. As the robustness of the model has become an important issue in control theory, the chapter discussed several approaches used to determine uncertainty associated with a neural model. It should be pointed out that the presented approaches are aimed to work in the context of nonlinear models. The chapter contains a detailed description of set-membership identification, model error modelling as well as the problem of determining statistical bounds for nonlinear models. In summary, the artificial neural networks represent a very attractive modelling tool. They seem to be a perfect fit for solving the problems related to the control system synthesis. The next chapter constitutes a review of different control strategies and roles neural networks can play in the framework of a control system design.

References

1. Ackley, D.H., Hinton, G.E., Sejnowski, T.J.: A learning algorithm for Boltzmann machines. Cogn. Sci. **9**, 147–169 (1985)
2. Anderson, S., Merrill, J.W.L., Port, R.: Dynamic speech categorization with recurrent networks. In: Touretzky, D., Hinton, G., Sejnowski, T. (eds.) Proceedings of the 1988 Connectionist Models Summer School (Pittsburg 1988), pp. 398–406. Morgan Kaufmann, San Mateo (1989)
3. Antonelo, E.A., Camponogara, E., Foss, B.: Echo state networks for data-driven downhole pressure estimation in gas-lift oil wells. Neural Netw. **85**, 106–117 (2017)
4. Atkinson, A.C., Donev, A.N., Tobias, R.D.: Optimum Experimental Designs, with SAS. Oxford University Press, Oxford (2007)
5. Auda, G., Kamel, M.: CMNN: cooperative modular neural networks for pattern recognition. Pattern Recognit. Lett. **18**, 1391–1398 (1997)
6. Ayoubi, M.: Fault diagnosis with dynamic neural structure and application to a turbo-charger. In: Proceedings of the International Symposium on Fault Detection Supervision and Safety for Technical Processes, SAFEPROCESS'94, Espoo, Finland, vol. 2, pp. 618–623 (1994)
7. Back, A.D., Tsoi, A.C.: FIR and IIR synapses, a new neural network architecture for time series modelling. Neural Comput. **3**, 375–385 (1991)
8. Badoni, M., Singh, B., Singh, A.: Implementation of echo-state network-based control for power quality improvement. IEEE Trans. Ind. Electron. **64**, 5576–5584 (2017)
9. Battlori, R., Laramee, C.B., Land, W., Schaffer, J.D.: Evolving spiking neural networks for robot control. Procedia Comput. Sci. **6**, 329–334 (2011)
10. Bianchi, F.M., Livi, L., Alippi, C.: Investigating echo-state networks dynamics by means of recurrence analysis. IEEE Trans. Neural Netw. Learn. Syst. **29**, 427–439 (2018)
11. Camacho, E.F., Bordóns, C.: Model Predictive Control, 2nd edn. Springer, London (2007)
12. Campolucci, P., Uncini, A., Piazza, F., Rao, B.D.: On-line learning algorithms for locally recurrent neural networks. IEEE Trans. Neural Netw. **10**, 253–271 (1999)
13. Chen, S., Billings, S.A.: Neural network for nonliner dynamic system modelling and identification. Int. J. Control **56**, 319–346 (1992)
14. Choi, B.B., Lawrence, C.: Inverse kinematics problem in robotics using neural network. Technical report 105869, NASA (1992)
15. Cybenko, G.: Approximation by superpositions of a sigmoidal function. Math. Control Signals Syst. **2**, 303–314 (1989)
16. Czajkowski, A., Patan, K., Szymański, M.: Application of the state space neural network to the fault tolerant control system of the PLC-controlled laboratory stand. Eng. Appl. Artif. Intell. **30**, 168–178 (2014)
17. Demuth, H., Beale, M.: Neural Network Toolbox for Use with MATLAB. The MathWorks Inc, Natick (1993)
18. Ding, L., Gustafsson, T., Johansson, A.: Model parameter estimation of simplified linear models for a continuous paper pulp degester. J. Process Control **17**, 115–127 (2007)
19. Eckhorn, R., Reitbock, H.J., Arndt, M., Dicke, P.: A neural network for feature linking via synchronous activity: results from cat visual cortex and from simulations. In: Cotterill, R.M.J. (ed.) Models of Brain Function, pp. 255–272. Cambridge University Press, Cambridge (1989)
20. Elman, J.L.: Finding structure in time. Cogn. Sci. **14**, 179–211 (1990)

21. Fahlman, S.E.: Fast learning variation on back-propagation: an empirical study. In: Touretzky, D., Hilton, G., Sejnowski, T. (eds.) Proceedings of the 1988 Connectionist Models Summer School (Pittsburg 1988), pp. 38–51. Morgan Kaufmann, San Mateo (1989)
22. Fasconi, P., Gori, M., Soda, G.: Local feedback multilayered networks. Neural Comput. **4**, 120–130 (1992)
23. Fedorov, V.V., Hackl, P.: Model-Oriented Design of Experiments. Lecture Notes in Statistics. Springer, New York (1997)
24. Ferrari, S., Stengel, R.F.: Smooth function approximation using neural networks. IEEE Trans. Neural Netw. **16**, 24–38 (2005)
25. Garzon, M., Botelho, F.: Dynamical approximation by recurrent neural networks. Neurocomputing **29**, 25–46 (1999)
26. Gerstner, W., Kistler, W.M.: Spiking Neuron Models. Single Neurons, Populations, Plasticity. Cambridge University Press, Cambridge (2002)
27. Girosi, J., Poggio, T.: Neural network and the best approximation property. Biol. Cybern. **63**, 169–176 (1990)
28. Gori, M., Bengio, Y., Mori, R.D.: BPS: a learning algorithm for capturing the dynamic nature of speech. In: International Joint Conference on Neural Networks, vol. II, pp. 417–423 (1989)
29. Gunnarson, S.: On some asymptotic uncertainty bounds in recursive least squares identification. IEEE Trans. Autom. Control **38**, 1685–1689 (1993)
30. Gupta, M.M., Jin, L., Homma, N.: Static and Dynamic Neural Networks. From Fundamentals to Advanced Theory. Wiley, New Jersey (2003)
31. Gupta, M.M., Rao, D.H.: Dynamic neural units with application to the control of unknown nonlinear systems. J. Intell. Fuzzy Syst. **1**, 73–92 (1993)
32. Hagan, M., Demuth, H.B., Beale, M.H.: Neural Network Design. PWS Publishing, Boston (1996)
33. Hagan, M.T., Menhaj, M.B.: Training feedforward networks with the Marquardt algorithm. IEEE Trans. Neural Netw. **5**, 989–993 (1994)
34. Haykin, S.: Neural Networks. A Comprehensive Foundation, 2nd edn. Prentice-Hall, New Jersey (1999)
35. Hertz, J., Krogh, A., Palmer, R.G.: Introduction to the Theory of Neural Computation. Addison-Wesley Publishing Company, Inc., Reading (1991)
36. Hinton, G.E., Osindero, S., Teh, Y.W.: A fast learning algorithm for deep belief nets. Neural Comput. **18**, 1527–1554 (2006)
37. Hinton, G.E., Salakhutdinov, R.R.: Reducing the dimensionality of data with neural networks. Science **313**, 504–507 (2006)
38. Hochreiter, S., Schmidhuber, J.: Long short-term memory. Neural Comput. **9**, 1735–1780 (1997)
39. Hopfield, J.J.: Neural networks as physical systems with emergent collective computational abilities. In: Proceedings of the National Academy of Sciences, vol. 79, pp. 2554–2558 (1982)
40. Hornik, K., Stinchcombe, M., White, H.: Multilayer feedforward networks are universal approximators. Neural Netw. **2**, 359–366 (1989)
41. Hunt, K.J., Sbarbaro, D., Zbikowski, R., Gathrop, P.J.: Neural networks for control systems – a survey. Automatica **28**, 1083–1112 (1992)
42. Isermann, R., Münchhof, M.: Neural networks and lookup tables for identification. In: Identification of Dynamic Systems. Springer, Berlin (2011)
43. Izhikevich, E.M.: Simple model of spiking neurons. IEEE Trans. Neural Netw. **14**, 1569–1572 (2003)
44. Jaeger, H.: The echo state approach to analysing and training recurrent neural networks. Technical report. GMD report 148, German National Research Center for Information Technology, Germany (2001)
45. Janczak, A.: Identification of Nonlinear Systems Using Neural Networks and Polynomial Models. A Block-Oriented Approach. Lecture Notes in Control and Information Sciences. Springer, Berlin (2005)

46. Jin, L., Nikiforuk, P.N., Gupta, M.M.: Approximation of discrete-time state-space trajectories using dynamic recurrent neural networks. IEEE Trans. Autom. Control **40**, 1266–1270 (1995)
47. Johnson, J.L.: Pulse-coupled neural nets: translation, rotation, scale, distortion, and intensity signal invariance for images. Appl. Opt. **33**(26), 6239–6253 (1994)
48. Johnson, J.L., Padgett, M.L.: PCNN models and applications. Neural Netw. **10**(3), 480–498 (1999)
49. Johnson, J.L., Ritter, D.: Observation of periodic waves in a pulse-coupled neural network. Opt. Lett. **18**(15), 1253–1255 (1993)
50. Jordan, M.I.: Attractor dynamic and parallelism in a connectionist sequential machine. In: Proceedings of the 8th Annual Conference of the Cognitive Science Society (Amherst, 1986), pp. 531–546. Erlbaum, Hillsdale (1986)
51. Jordan, M.I., Jacobs, R.A.: Supervised learning and systems with excess degrees of freedom. In: Touretzky, D.S. (ed.) Advances in Neural Information Processing Systems II (Denver 1989), pp. 324–331. Morgan Kaufmann, San Mateo (1990)
52. Kohonen, T.: Self-organization and Associative Memory. Springer, Berlin (1984)
53. Kohonen, T.: Self-organizing Maps. Springer, Berlin (2001)
54. Korbicz, J., Kościelny, J., Kowalczuk, Z., Cholewa, W. (eds.) Fault Diagnosis. Models, Artificial Intelligence, Applications. Springer, Berlin (2004)
55. Kuschewski, J.G., Hui, S., Żak, S.: Application of feedforward neural network to dynamical system identification and control. IEEE Trans. Neural Netw. **1**, 37–49 (1993)
56. Ławryńczuk, M.: Computationally Efficient Model Predictive Control Algorithms. A Neural Network Approach. Studies in Systems, Decision and Control, vol. 3. Springer, Switzerland (2014)
57. Leshno, M., Lin, V., Pinkus, A., Schoken, S.: Multilayer feedforward networks with a nonpolynomial activation function can approximate any function. Neural Netw. **6**, 861–867 (1993)
58. Lindblad, T., Kinser, J.M.: Image Processing Using Pulse-Coupled Neural Networks. Springer, London (1998)
59. Ma, Y., Zhan, K., Wang, Z.: Applications of Pulse-Coupled Neural Networks. Springer, Berlin (2010)
60. Maass, W.: Networks of spiking neurons: the third generation of neural network models. Neural Netw. **10**, 1659–1671 (1997)
61. Maass, W., Natschlaeger, T., Markram, H.: Real-time computing without stable states: a new framework for neural computation based on perturbations. Neural Comput. **14**, 2531–2560 (2002)
62. Marciniak, A., Korbicz, J.: Diagnosis system based on multiple neural classifiers. Bull. Pol. Acad. Sci. Tech. Sci. **49**, 681–701 (2001)
63. McCulloch, W.S., Pitts, W.: A logical calculus of ideas immanent in nervous activity. Bull. Math. Biophys. **5**, 115–133 (1943)
64. Milanese, M.: Set membership identification of nonlinear systems. Automatica **40**, 957–975 (2004)
65. Mozer, M.C.: A focused backpropagation algorithm for temporal pattern recognition. Complex Syst. **3**, 349–381 (1989)
66. Narendra, K.S., Parthasarathy, K.: Identification and control of dynamical systems using neural networks. IEEE Trans. Neural Netw. **1**, 12–18 (1990)
67. Nelles, O.: Nonlinear System Identification. From Classical Approaches to Neural Networks and Fuzzy Models. Springer, Berlin (2001)
68. Nørgaard, M., Ravn, O., Poulsen, N., Hansen, L.: Networks for Modelling and Control of Dynamic Systems. Springer, London (2000)
69. Parlos, A.G., Chong, K.T., Atiya, A.F.: Application of the recurrent multilayer perceptron in modelling complex process dynamics. IEEE Trans. Neural Netw. **5**, 255–266 (1994)
70. Patan, K.: Robust fault diagnosis in catalytic cracking converter using artificial neural networks. In: Proceedings of the 16th IFAC World Congress, 3–8 July, Prague, Czech Republic (2005). Published on CD-ROM

71. Patan, K.: Approximation of state-space trajectories by locally recurrent globally feed-forward neural networks. Neural Netw. **21**, 59–63 (2008)
72. Patan, K.: Artificial Neural Networks for the Modelling and Fault Diagnosis of Technical Processes. Lecture Notes in Control and Information Sciences. Springer, Berlin (2008)
73. Patan, K.: Two stage neural network modelling for robust model predictive control. ISA Trans. **72**, 56–65 (2018)
74. Patan, K., Korbicz, J., Głowacki, G.: DC motor fault diagnosis by means of artificial neural networks. In: Proceedings of the 4th International Conference on Informatics in Control, Automation and Robotics, ICINCO 2007, Angers, France, 9–12 May 2007. Published on CD-ROM
75. Patan, K., Parisini, T.: Stochastic learning methods for dynamic neural networks: simulated and real-data comparisons. In: Proceedings of the 2002 American Control Conference (IEEE Cat. No.CH37301), vol. 4, pp. 2577–2582 (2002)
76. Patan, K., Parisini, T.: Identification of neural dynamic models for fault detection and isolation: the case of a real sugar evaporation process. J. Process Control **15**, 67–79 (2005)
77. Patan, K., Patan, M.: Optimal training sequences for locally recurrent neural network. Lect. Notes Comput. Sci. **5768**, 80–89 (2009)
78. Patan, K., Patan, M., Kowalów, D.: Optimal sensor selection for model identification in iterative learning control of spatio-temporal systems. In: 55th IEEE Conference on Decision and Control (CDC) (2016)
79. Patan, M.: Distributed scheduling of sensor networks for identification of spatio-temporal processes. Appl. Math. Comput. Sci. **22**(2), 299–311 (2012)
80. Patan, M.: Sensor Networks Scheduling for Identification of Distributed Systems. Lecture Notes in Control and Information Sciences, vol. 425. Springer, Berlin (2012)
81. Patan, M., Bogacka, B.: Optimum group designs for random-effects nonlinear dynamic processes. Chemom. Intell. Lab. Syst. **101**, 73–86 (2010)
82. Pearlmutter, B.A.: Learning state space trajectories in recurrent neural networks. In: International Joint Conference on Neural Networks (Washington 1989), vol. II, pp. 365–372. IEEE, New York (1989)
83. Pham, D.T., Xing, L.: Neural Networks for Identification. Prediction and Control. Springer, Berlin (1995)
84. Plaut, D., Nowlan, S., Hinton, G.: Experiments of learning by back propagation. Technical report CMU-CS-86-126, Department of Computer Science, Carnegie Melon University, Pittsburg, PA (1986)
85. Poddar, P., Unnikrishnan, K.P.: Memory neuron networks: a prolegomenon. Technical report GMR-7493, General Motors Research Laboratories (1991)
86. Psaltis, D., Sideris, A., Yamamura, A.A.: A multilayered neural network controller. IEEE Control Syst. Mag. **8**(2), 17–21 (1988)
87. Quinn, S.L., Harris, T.J., Bacon, D.W.: Accounting for uncertainty in control-relevant statistics. J. Process Control **15**, 675–690 (2005)
88. Reinelt, W., Garulli, A., Ljung, L.: Comparing different approaches to model error modeling in robust identification. Automatica **38**, 787–803 (2002)
89. Rojas, R.: Neural Networks. A Systematic Introduction. Springer, Berlin (1996)
90. Rosenblatt, F.: Principles of Neurodynamics: Perceptrons and the Theory of Brain Mechanisms. Spartan Books, Washington (1962)
91. Rumelhart, D.E., Hinton, G.E., Williams, R.J.: Learning internal representations by error propagation. Parallel Distributed Processing, vol. I (1986)
92. Rumelhart, D.E., Hinton, G.E., Williams, R.J.: Learning representations by back-propagating errors. Nature **323**, 533–536 (1986)
93. Sastry, P.S., Santharam, G., Unnikrishnan, K.P.: Memory neuron networks for identification and control of dynamical systems. IEEE Trans. Neural Netw. **5**, 306–319 (1994)
94. Smolensky, P.: Information processing in dynamical systems: foundations of harmony theory. In: Rumelhart, D.E., McLelland, J.L. (eds.) Parallel Distributed Processing: Explorations in the Microstructure of Cognitio, pp. 194–281. MIT Press (1986)

95. Sollich, P., Krogh, A.: Learning with ensembles: how over-fitting can be useful. In: Advances in Neural Information Processing System. Proceedings of the 1996 Conference, vol. 9, pp. 190–196 (1996)
96. Sontag, E.: Feedback stabilization using two-hidden-layer nets. IEEE Trans. Neural Netw. **3**, 981–990 (1992)
97. Sørensen, O.: Neural networks performing system identification for control applications. In: Proceedings of the 3rd International Conference on Artificial Neural Networks, Brighton, UK, pp. 172–176 (1993)
98. Specht, D.F.: Probabilistic neural networks. Neural Netw. **3**, 109–118 (1990)
99. Stornetta, W.S., Hogg, T., Hubermann, B.A.: A dynamic approach to temporal pattern processing. In: Anderson, D.Z. (ed.) Neural Information Processing Systems, pp. 750–759. American Institute of Physics, New York (1988)
100. Tsoi, A.C., Back, A.D.: Locally recurrent globally feedforward networks: a critical review of architectures. IEEE Trans. Neural Netw. **5**, 229–239 (1994)
101. Ucinski, D.: Optimal Measurement Methods for Distributed Parameter System Identification. CRC Press, Boca Raton (2004)
102. Uciński, D.: Sensor network scheduling for identification of spatially distributed processes. Appl. Math. Comput. Sci. **22**(1), 25–40 (2012)
103. Walter, E., Pronzato, L.: Identification of Parametric Models from Experimental Data. Springer, London (1997)
104. Wang, X., Hou, Z.G., Lv, F., Tan, M., Wang, Y.: Mobile robots' modular navigation controller using spiking neural networks. Neurocomputing **134**, 230–238 (2014)
105. Warwick, K., Kambhampati, C., Parks, P., Mason, J.: Dynamic systems in neural networks. In: Hunt, K.J., Irwin, G.R., Warwick, K. (eds.) Neural Network Engineering in Dynamic Control Systems, pp. 27–41. Springer, Berlin (1995)
106. Werbos, P.J.: Beyond regression: new tools for prediction and analysis in the behavioral sciences. Ph.D. thesis, Harvard University (1974)
107. Widrow, B.: Generalization and information storage in networks of adaline neurons. In: Yovits, M., Jacobi, G.T., Goldstein, G. (eds.) Self-organizing Systems 1962 (Chicago 1962), pp. 435–461. Spartan, Washington (1962)
108. Widrow, B., Hoff, M.E.: Adaptive switching circuit. In: 1960 IRE WESCON Convention Record, Part 4, pp. 96–104. IRE, New York (1960)
109. Wiklendt, L., Chalup, S., Middleton, R.: A small spiking neural network with LQR control applied to the acrobot. Neural Comput. Appl. **18**, 369–375 (2009)
110. Williams, R.J., Zipser, D.: Experimental analysis of the real-time recurrent learning algorithm. Connect. Sci. **1**, 87–111 (1989)
111. Williams, R.J., Zipser, D.: A learning algorithm for continually running fully recurrent neural networks. Neural Comput. **1**, 270–289 (1989)
112. Xu, L., Krzyzak, A., Suen, C.: Methods for combining multiple classifiers and their applications to handwriting recognition. IEEE Trans. Syst. Man Cybern. **22**, 418–435 (1992)
113. Zamarreno, J.M., Vega, P.: State space neural network. Properties and application. Neural Netw. **11**, 1099–1112 (1998)
114. Żurada, J.M.: Lambda learning rule for feedforward neural networks. In: Proceedings of the International Conference on Neural Networks, San Francisco, USA, March 28–April 1, pp. 1808–1811 (1993)

Chapter 3
Robust and Fault-Tolerant Control

3.1 The Context of Control Systems

It can be said that the first control systems were developed over two thousand years ago, however, the control theory in the modern sense can be traced back to the developments in the 1950s and 1960s when early solutions for optimal control were proposed. Currently, control systems are commonly used not only in industrial applications but also in our daily life, e.g. temperature control in refrigerators, central heating systems or ovens; cruise control in cars, etc.

Process control consists in manipulating variables, controlled variables and plants/processes. Usually, the manipulating and controlled variables correspond to the plant inputs and outputs, respectively. The main objective of a control system is to make the process under control behave in a desired way [23]. This can be done by properly manipulating the plant input. There are two paradigms of the control systems design: open-loop control and closed-loop control [19]. In the systems belonging to the former, the control signal determined by the controller is independent of the measurable variable (the output) of the process. As an example of an open-loop control let us consider an irrigation sprinkler system (see Fig. 3.1). The timer plays the role of the controller and the sprinkler is the process with irrigation being the controlled variable. The control system is programmed to switch on/off at set times. A significant disadvantage of such a solution is that the system will activate according to the schedule even in the case of rain, thus wasting water. In turn, in the closed-loop control, the control law is determined based on the desired and actual process output values. In the context of the irrigation control system, the closed-loop can be realized using a soil moisture sensor. In such a setting, the current value of the soil moisture is fed back to the controller and compared with the desired value. If the soil is too dry, the sprinkler system will activate according to the schedule, otherwise the system will remain in the off state.

Consider yet another example of a feedback control system — this time the flight control system depicted in Fig. 3.2. The autopilot is used to control the trajectory of an airplane in a situation, when a constant manual control of a human operator is not

© Springer Nature Switzerland AG 2019

K. Patan, *Robust and Fault-Tolerant Control*, Studies in Systems,
Decision and Control 197, https://doi.org/10.1007/978-3-030-11869-3_3

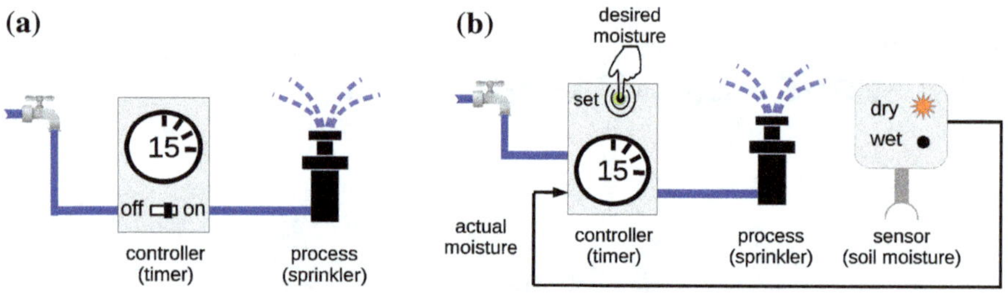

Fig. 3.1 Open-loop (**a**) versus closed-loop control (**b**)

Fig. 3.2 An example of an automatic control system

required. The objective of the control system is to keep the desired flight parameters regardless of the disturbances influencing the behaviour of the aircraft, e.g. a wind, a turbulence, etc. Given that, there are two manipulating variables: the throttle position and the elevator position. Let us assume that the controlled variables are the speed and altitude of the aircraft. These can be measured by suitable sensors and are compared with the desired set-up values. The calculated differences (regulation errors) drive the controller with the aim to achieve the desired control objectives.

A general scheme of an automatic control system is depicted in Fig. 3.3. The controller derives the control $u(k)$ based on the regulation error $e(k)$ defined as the difference between the reference signal $r(k)$ and the plant output $y(k)$. The signal $y(k)$ is acquired from the plant using a suitable sensor. The control signal drives

Fig. 3.3 A general scheme of the closed-loop control system

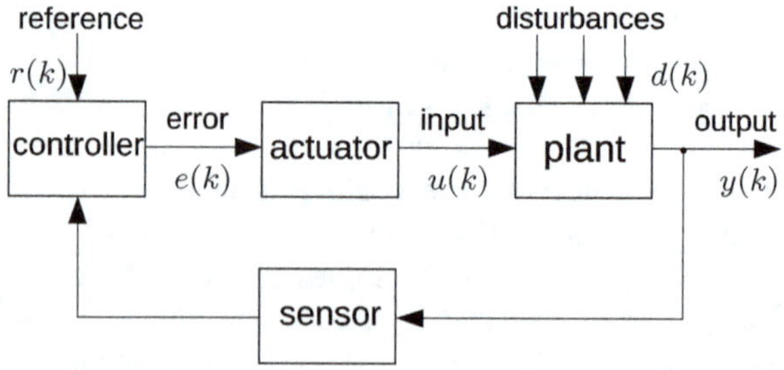

the actuator which acts upon the plant. The control problem can be formulated in a number of different ways [23]:

- to derive a control signal $u(k)$ in such a way as to counteract the effect of disturbance $d(k)$ influencing the plant,
- to derive a control signal $u(k)$ in such a way as to keep the plant output the closest to a given reference $r(k)$.

In both cases the control error $e(k)$ should be as small as possible. Typically, the controller synthesis is based on a mathematical model of a plant, most frequently using linear models [23]. A mathematical model of the process can be constructed using the available data recorded in the plant. This constitutes the essence of the identification and parameter estimation problem. Process control can be difficult when the process itself, as represented by the model, is difficult to control, or when the information available to the controller is imprecise or incorrect. The following lists certain features that can make a real-life process difficult to control [17]:

1. the order of the process is not known,
2. the process is be dynamic but the operational data available on the input and output are not rich and do not reveal the important process characteristics,
3. the time delay between the input and the output is not fixed or known,
4. the process is open-loop unstable,
5. the process is nonminimum-phase in nature.

The first three features make the correct identification and parameter estimation of the process difficult. The last three features demand particular attention in terms of the choice of the cost criterion that the control law must optimize. Additionally, the difficulties caused by unknown disturbances $d(k)$ acting on the process can also be observed. All these factors could make the identification process difficult or even impossible to be properly finalised.

3.1.1 Control Based on Neural Networks

To date, a large number of control structures based on neural networks models have been developed and applied [13, 15, 22, 29]. According to the classification presented in Fig. 3.4, the approach to control problems using neural networks is a subclass of a well-understood sub-field called intelligent control. Neural networks are especially helpful in the cases, when the analytical model of the controlled plant is hard or impossible to obtain or when the controlled plant is highly nonlinear in its nature. The ability of neural networks to represent nonlinear systems is perhaps the most important in this context. In spite of the fact that one is faced with an abundance of classical synthesis methods of nonlinear controllers for specific classes of nonlinear systems i.e. phase-plane methods, linearization techniques or describing functions, the ability of neural networks to deal with nonlinear mappings is the feature to be most readily exploited in a synthesis of nonlinear controllers. Below, we briefly discuss the most popular approaches to control that make use of neural networks.

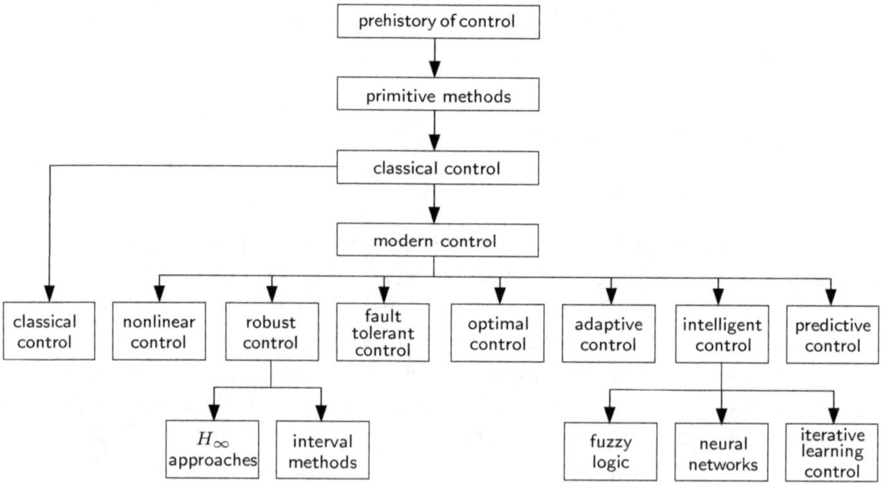

Fig. 3.4 A classification of the control strategies [21]. Reproduced by courtesy of Hyo-Sung Ahn

Direct Inverse Control

Direct inverse control uses an inverse model of a plant. Assume that the controlled plant is described by the following difference equation:

$$y(k+1) = f(y(k), \ldots, y(k-n_a+1), u(k), \ldots, u(k-n_b+1)), \tag{3.1}$$

where n_a and n_b are number of output and input delayed signals respectively. The neural network inverse model is represented as [22, 24]:

$$\hat{u}(k) = f^{-1}(y(k+1), y(k), \ldots, y(k-n_a+1), u(k), \ldots, u(k-n_b+1)). \tag{3.2}$$

However, the inverse model (3.2) requires knowledge about the future value of the plant output $y(k+1)$. To deal with this problem, it is assumed that $y(k+1)$ can be replaced by the reference signal $r(k+1)$ assumed to be known at the time instant k. In such a situation, (3.2) can be rewritten as follows:

$$\hat{u}(k) = f^{-1}(r(k+1), y(k), \ldots, y(k-n_a+1), u(k), \ldots, u(k-n_b+1)) \tag{3.3}$$

leading to the scheme presented in Fig. 3.5. The key problem here is to properly design the neural network controller. The problem of training the neural network in order to get the inverse of the plant is discussed in Chap. 2, Sect. 2.4.2. In short, the distinct characteristics of the direct inverse control can be listed as follows:

- intuitive and simple to implement,
- does not work in a case of a plant with an unstable inverse,
- problems with not well-damped inverse models,
- lack of robustness due to absence of feedback.

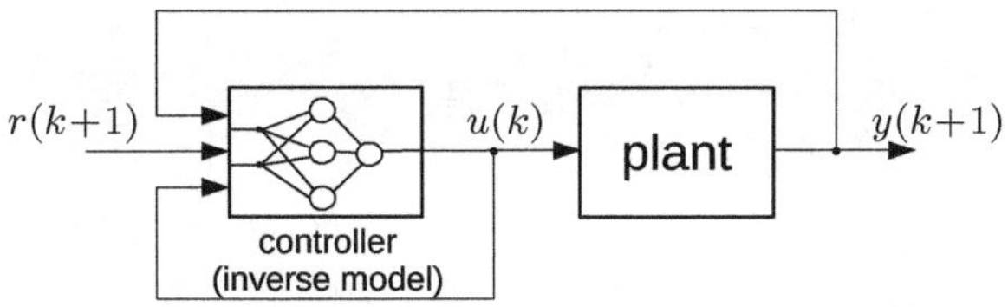

Fig. 3.5 Direct control

Model Reference Control

Model reference control is also referred to as model reference adaptive control (MRAC). The desired performance of the control system is specified by a stable reference model. In many cases, the reference model is defined as the input-output model, where $r(k)$ is the input and $y_m(k)$ is the output. The role of the controller is to match the plant output $y(k)$ to the output of the reference model $y_m(k)$ as accurately as possible, e.g.

$$\lim_{k\to\infty} \|y_m(k) - y(k)\| \leq \epsilon, \tag{3.4}$$

where $\epsilon \geq 0$. In the case of a nonlinear system, one can use a dynamic neural network to design the controller. In that case, the error $\varepsilon(k)$, defined as the difference between the output of the reference model $y_m(k)$ and output of the plant $y(k)$, is used to train the neural network controller using a suitably selected training algorithm [12, 13, 22]. If the reference model is defined as the identity mapping, the model reference control leads to a direct inverse control, when the controller parameters are updated on-line (Fig. 3.6).

Internal Model Control

Internal model control (IMC) relies on the internal model principle, which states that accurate control can be achieved only if the control systems encapsulate a representation of the controlled process. Assuming this, the control system can be said to consists of both the model of the plant as well as the inverse model. In the case of nonlinear systems, both models can be realized by means of neural networks. The first neural network, playing the role of the forward model, is placed in parallel with the real plant [13, 31]. The difference between the system output $y(k)$ and the model

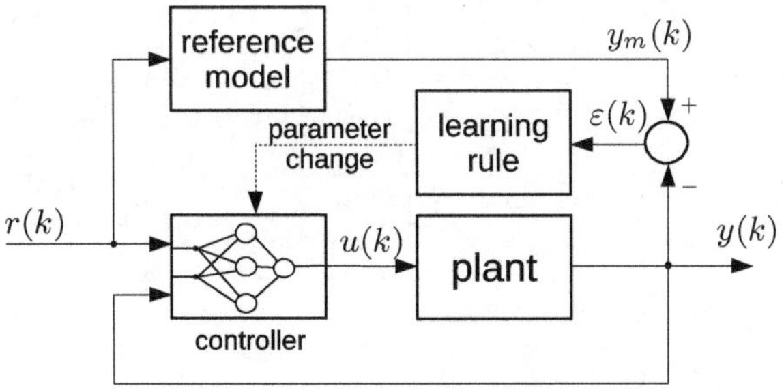

Fig. 3.6 Model reference control

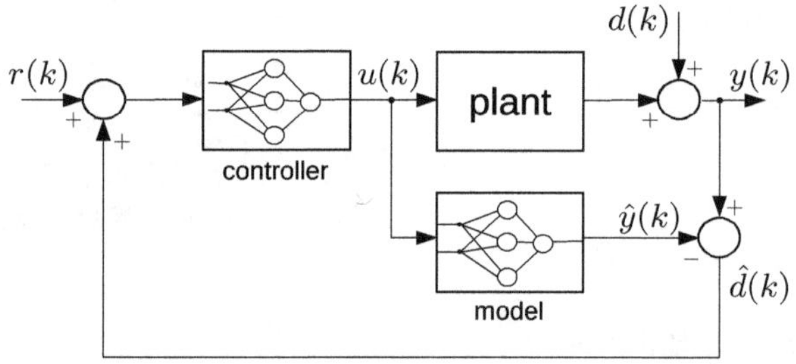

Fig. 3.7 Internal model control

output $\hat{y}(k)$ is the estimated effect of the disturbance $d(k)$ that influences the plant. It is denoted by $\hat{d}(k)$. The signal $\hat{d}(k)$ constitutes the feedback and it is processed in the forward path by the second neural network acting as the controller. Taking into account the properties of IMC, the controller is related to the plant inverse. It should be noted that the implementation of IMC is limited to the open-loop stable processes. Despite this, IMC is widely used in process control. Even if the true inverse of the system does not exists, neural networks can provide useful approximations making IMC useful in terms of practical realizations (Fig. 3.7).

Feed-Forward Control

Applying the inverse models for feedback control leads to the so-called dead-beat control, which is problematic in many cases. Dead-beat control usually requires a large control action, which may saturate actuators. Moreover, the observed overshoot is usually very high, and the control signal can be expensive to generate. The idea of feed-forward control is to combine a feedback controller with a feed-forward one (Fig. 3.8) [15, 24]. The augmented control signal is as follows:

$$u(k) = u^{ff}(k) + u^{fb}(k), \tag{3.5}$$

where $u^{fb}(k)$ is the signal provided by the feedback controller, and $u^{ff}(k)$ is the control component generated by the feed-forward controller. In many cases one can use the existing feedback control system with an already tuned PID controller designed for the purpose of stabilizing the plant. In such a case, feed-forward control can be used very successfully to improve a control system response to disturbances.

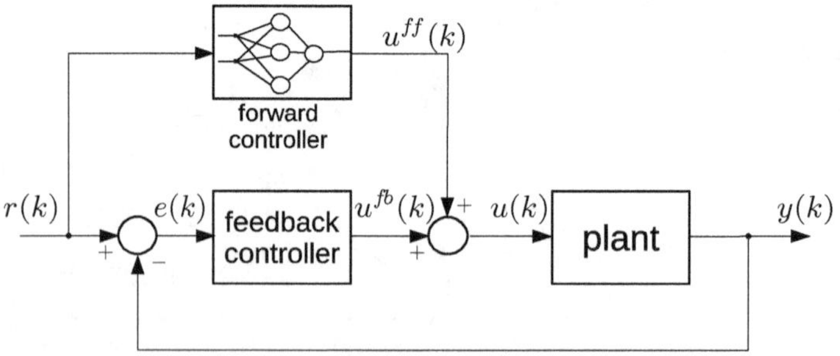

Fig. 3.8 Feed-forward control

For example, the set-point change can act as a disturbance on the feedback loop. The feed-forward controller can react before the effect of the disturbance shows up in the plant output. In fact, the controller is in the form of an inverse model, which is used for providing a feed-forward signal directly from the reference:

$$u^{ff}(k) = g^{-1}(r(k+1), r(k), \ldots, r(k-n_a+1), u^{ff}(k-1), \ldots, u^{ff}(k-n_b+1)), \tag{3.6}$$

where g^{-1} represents the inverse function n_a and n_b represent the delays. A simpler nonlinear finite impulse response model can also be used:

$$u^{ff}(k) = g^{-1}(r(k+1), r(k), \ldots, r(k-n_a+1)). \tag{3.7}$$

As a result, the control system is capable of fast tracking the reference.

Predictive Control

Model predictive control is a specific control strategy that uses a model of the process to derive the control signal by minimizing some objective function over a finite receding horizon. The control sequence is selected as to minimize, at each iteration, a quadratic criterion:

$$J = \sum_{i=N_1}^{N_2} (r(k+i) - \hat{y}(k+i))^2 + \rho \sum_{i=1}^{N_u} (\Delta u(k+i-1))^2 \tag{3.8}$$

with respect to N_u future controls, and subject to constraints. Here, the constants N_1 and N_2 define the horizon, over which the control sequence is derived. The parameter ρ is a penalty imposed on the control changes. The scheme of model predictive control is shown in Fig. 3.9, where a neural network is designed to mimic the future response of the plant required for cost optimization [16]. The neural network model is constructed using the past and current outputs as well as the future control signals. These are calculated through the minimization of the cost function and taking into account constraints. There are a number of advantages of model predictive control:

- it is very easy to consider various constraints imposed on process variables,
- it can be used to control plants with time delay,
- it can deal with disturbances with known characteristics.

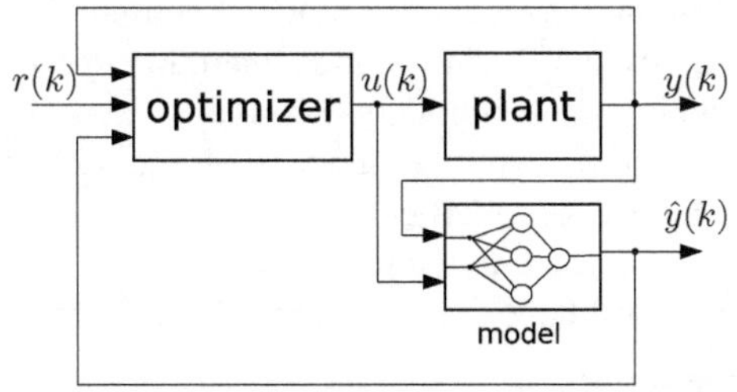

Fig. 3.9 Predictive control

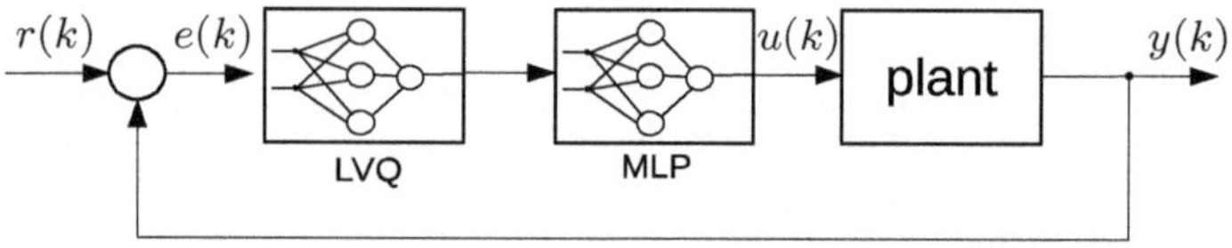

Fig. 3.10 Optimal control

The main disadvantage of such kind of control is that optimization can be troublesome and time-consuming, especially when dealing with nonlinear systems.

Optimal Control

In this scheme, the state space is partitioned into the feature space. Each feature corresponds to a specific control situation. The controller consists of two components. The first one quantizes the state-space into elementary hypercubes, in which the control signal is assumed to be constant. Then, each elementary hypercube needs to be classified in order to give a suitable control action. Generally, the time-optimal surface is nonlinear. Therefore, it is necessary to apply the methods allowing one to deal with the nonlinear problems. Figure 3.10 shows the scheme of optimal control constructed using neural networks. The first presented neural network is an example of a learning vector quantization structure, which is able to quantize the input space. The second neural network can be a simple feed-forward network able to classify the patterns [13]. The switching surface is not known a priori but it is defined implicitly by a training algorithm applied to a set of points in the state-space whose optimal control action is known.

3.2 Robust Control

Robust control is a branch of control theory that explicitly deals with system uncertainty and how uncertainty affects the analysis and design of control systems. We can say that a control system is robust, if it is insensitive to differences between the controlled system and the model used in the context of the controller design. These differences are referred to as model mismatch or model uncertainty. The term robust control first appeared in the title of a conference paper by Davison [6] and in a journal paper by Pearson and Staats in 1974 [28]. However, the first attempts to develop robust control schemes started 10 years earlier with the pioneering works of Zames in 1963 [45] and Kalman in 1964 [14]. Zames introduced the concept of the small-gain principle, which plays a crucial role in robust stability. In turn, Kalman showed that optimal linear quadratic (LQ) state feedback controller possesses strong robustness properties.

According to the paper by Safonov [36], the milestones of the early development of robust control can be formulated as follows:

1975–1977: Diagonally structured uncertainty. In 1975, Poh Wong formulated the multivariate stability margin problem in the form of a matrix of simultaneously varying uncertain real gains, which he analysed using Lyapunov methods [43]. This theory was extended to frequency-dependent complex uncertainties [37]. In 1978 Doyle demonstrated poor robustness of linear quadratic Gaussian (LQG) controllers [8].

1977: Robustness thesis. Safonov in his Ph.D. Thesis [34], introduced certain fundamental robustness concepts, and developed the forerunners of the singular-value, as well as integral-quadratic constraint (IQC) methods.

1978–1979: Singular values. Although at that time, the singular-value stability conditions had not been fully investigated yet in the context of control theory, a very similar representation of small-gain conditions was proposed for nonlinear stability analysis by Sandberg in 1964 [38]. The first published work on singular-value robustness was mentioned at the 1978 Allerton Conference during a special session.

1980–1982: Conservativeness, K_m and μ notation. In the early 1980s, several authors published papers on the role of diagonal scaling in singular value robustness tests [1, 30]. The diagonally-perturbed multivariate stability margin, known also as excess stability margin, was labeled as K_m [35]. Subsequently, Doyle [9] defined the structural singular value μ, thus developing μ-synthesis for robust controllers.

1980: Adaptive control counterexamples. Rohrs and colleagues [32, 33] showed the standard adaptive-control algorithms to have vanishingly small robustness, thus causing the turmoil of the robustness revolution spread to adaptive control.

1967–1989: H_∞ control. The term H_∞ was introduced by Zames [46] as a solution of sensitivity optimization problem. Later on, Doyle and colleagues [7] developed a state-space solution of H_∞ control synthesis.

1982: Performance robustness. The small-gain stability theorem can be interpreted as a performance robustness theorem. In this context, it is possible to include performance specifications into multivariate stability margin analysis by fictitious uncertainties [9].

1987: The first book on robust control. In 1987, Bruce Francis published a book entitled *A Course in H-infinity Control Theory* [10].

1988: Robust Control Toolbox for MATLAB. Owing to this engineering software suite, the robust control techniques become widely accessible to the engineers and practitioners.

3.2.1 Uncertainty Description

The possible sources of model uncertainty may be split into three categories [40]:

1. **Parametric uncertainty** – the structure of the model is known, however some of the parameters are unknown.

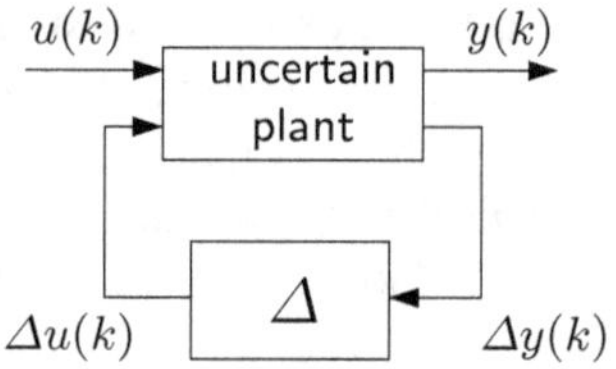

Fig. 3.11 The idea of structural uncertainty

2. **Neglected and unmodelled dynamics uncertainty** – the model is not accurate because it is not able to catch dynamics of the plant perfectly.
3. **Lumped uncertainty** – this kind of uncertainty is a combination of the other two listed above.

The notion of parametric uncertainty is used under the assumption that the structure of the model is known with uncertainty located in model parameters. It is assumed that each uncertain parameter is bounded in some region as $p \in [p_{min}, p_{max}]$. The possible set of values of the parameter p can be defined as follows:

$$p = \bar{p}(1 + r_p \Delta), \tag{3.9}$$

where $\bar{p}$ represents the mean value of p, Δ is any scalar satisfying $|\Delta| < 1$ and r_p stands for the relative uncertainty in the parameter p calculated as:

$$r_p = \frac{p_{max} - p_{min}}{p_{max} + p_{min}}. \tag{3.10}$$

Parametric uncertainty is alternatively known as structured uncertainty, because it models the uncertainty in a structured manner [20, 42]. The general idea of structural uncertainty is illustrated in Fig. 3.11. The operator Δ is the block diagonal

$$\Delta = \text{diag}\{\Delta_1, \ldots, \Delta_r\}, \tag{3.11}$$

where Δ_i represents a specific source of uncertainty. Such representation of uncertainty is frequently used when dealing with linear systems. The common assumption is that Δ is a stable system. As a direct consequence of such assumption we get that uncertainty satisfies the following condition [20]:

$$\|\Delta\|_\infty \leqslant 1. \tag{3.12}$$

The design process of the controller boils down to solving the H_∞ problem.

In turn, unmodelled dynamics uncertainty is somewhat less precise and thus more difficult to quantify. One of the possible solutions is to represent this kind of uncertainty in the frequency domain using the so-called multiplicative uncertainty, which — for the SISO linear systems — has the form:

$$G_p(s) = G(s)(1 + w_I(s)\Delta_I(s)), \quad \forall \omega\, |\Delta_I(j\omega)| \le 1, \tag{3.13}$$

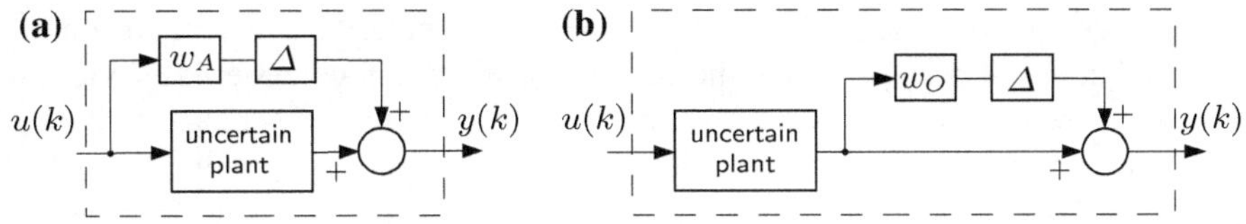

Fig. 3.12 Examples of unstructured uncertainty

where $\Delta_I(s)$ is any stable transfer function with the magnitude not greater than one for any frequency, $w_I(s)$ is a scalar weight and $G(s)$ is the transfer function of the nominal plant.

Lumped dynamics uncertainty is sometimes referred to as unstructured uncertainty. It is a combination of parametric uncertainty and unmodelled dynamics. In this case, uncertainty represents one or several sources of parametric or unmodelled dynamics uncertainty combined into a single lumped perturbation of a chosen structure. The examples of unstructured uncertainty are shown in Fig. 3.12 [40]. Figure 3.12a presents the additive uncertainty. Assuming that the uncertain plant is represent by a transfer function $G(s)$, the additive uncertainty form yields:

$$G_p(s) = G(s) + w_A \Delta, \tag{3.14}$$

where w_A is a scalar weight and Δ represents a perturbation matrix satisfying

$$\forall \omega \quad \sigma_{max}(\Delta(j\omega)) \leq 1, \tag{3.15}$$

where ω represents a frequency and σ_{max} stands for the maximum singular value. In turn, Fig. 3.12b illustrates multiplicative output uncertainty. In this case, the plant is represented as follows:

$$G_p(s) = G(s)(1 + w_O \Delta), \tag{3.16}$$

where w_O is the output scalar weight. Unstructured uncertainty forms are also very useful when dealing with linear systems. However, for nonlinear systems that kind of uncertainty modelling is insufficient.

Let us assume that a nonlinear plant is represented by the following state equation:

$$\boldsymbol{x}(k) = f(\boldsymbol{x}(k), \boldsymbol{u}(k)) \tag{3.17}$$

can be represented by a set of linear time varying models:

$$\mathscr{M} : \boldsymbol{x}(k+1) = \boldsymbol{A}(k)\boldsymbol{x}(k) + \boldsymbol{B}(k)\boldsymbol{u}(k), \tag{3.18}$$

where $[\boldsymbol{A}(k) \;\; \boldsymbol{B}(k)] \in \Omega$. The set Ω is the convex hull of the extreme models:

$$\Omega = \mathrm{Co}\{[\boldsymbol{A}_1 \;\; \boldsymbol{B}_1], \ldots, [\boldsymbol{A}_r \;\; \boldsymbol{B}_r]\}. \tag{3.19}$$

The number of models can be determined using, for example, the number of operating points of the nonlinear plant. As the number of the extreme models is finite, the convex hull is a polytope. Therefore, such kind of uncertainty is called polytopic uncertainty. The main advantage of the polytopic representation is that the robust control synthesis can be carried out using linear matrix inequalities (LMIs). However, polytopic uncertainty is a conservative approach to modelling a nonlinear system (3.17) [2].

Global uncertainty seems to be the uncertainty description that is appropriate for nonlinear systems modelling [4]. Let us consider a nonlinear system represented as follows:

$$\boldsymbol{x}(k+1) = f(\boldsymbol{x}(k), \boldsymbol{u}(k)) + \boldsymbol{w}(k), \tag{3.20}$$

where $\boldsymbol{w}(k) \in \mathcal{W}$ is the unknown disturbance. In this case, it is assumed that bounds imposed on the disturbance are known and $\mathcal{W}$ is a compact set. Global uncertainties can be considered to be more like disturbances than uncertainties, since they are considered in the system as external perturbations. The additive term $w(k)$ can be realized as a function of past plant inputs and outputs. If the process variables are bounded, the global uncertainties can also be bounded. In order to determine $w(k)$, the robust identification methods discussed in Chap. 2, Sect. 2.5 can be made use of.

3.3 Fault-Tolerant Control

Owing to the technological developments, modern industrial installations and control systems designed for them are increasingly more complex and are composed of ever growing number of interacting components and devices. The industrial plants are therefore becoming more vulnerable to non-permitted deviations of characteristic properties of the plant or unexpected changes of the system variables representing deviations from the acceptable/usual/normal behaviour. According to the IFAC Technical Committee SAFEPROCESS, such an non-permitted deviation is called a fault. Let us consider a general scheme of closed-loop control as shown in Fig. 3.13. A plant can be viewed as a set of interconnected elements including actuators, plant

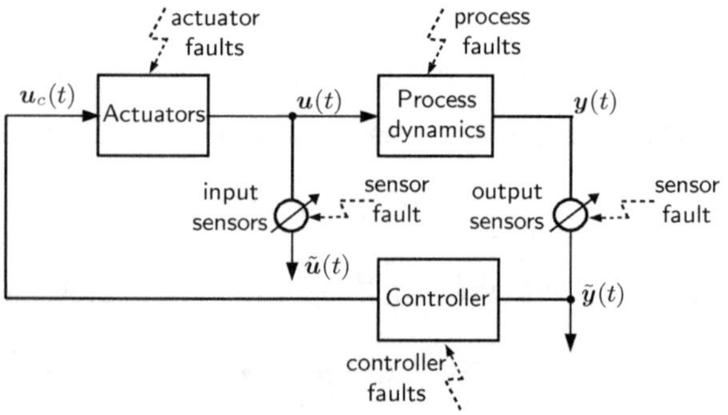

Fig. 3.13 Possible fault locations in the control loop

dynamics and sensors. Each element can be subject to a fault, so we can distinguish actuator faults, process faults and sensor faults. Moreover, faults can also occur in the controller itself. According to [11], actuator, process and sensors faults together can account for up to 60% of industrial control problems encountered. Moreover, faults in these specific parts may develop into a failure of the entire system. As most of modern control systems work in the closed-loop control, the fault effect can be amplified leading to the damage of the specific part of the plant. On the other hand, the closed-loop control might be able to compensate the fault effect and hide faults from being observed. Sooner or later a damaged component can cause an unexpected emergency shut down of the system. Such problems concern not only high-risk systems, such as nuclear plants, chemical systems or aircraft, but also a large number of modern industrial installations. Early fault detection and proper exploitation of the existing hardware or software redundancy can provide a way to avoid a shut-down or even a failure of the system. Such undesirable events are strongly related to the financial losses of companies and serious injuries or even death of personnel operating the specific system. In safety-critical systems, the requirement for a safe back-up in the case of failures is always present. FTC systems can be a more effective alternative in terms of providing the system back-up than the equipment redundancy, e.g. duplication or even triplication of actuators. The demand of fault-tolerant control comes also from economics. The costs of lost production due to a fault can be enormous, as can be the costs of unnecessary energy consumption. From that point of view, there is a strong encouragement to prolong the operation of a production plant despite faults until a scheduled maintenance period. The above-mentioned arguments constitute a strong motivation for the development of FTC systems. That is the reason why nowadays no efforts are being spared in designing control systems that are able to maintain the current performance of the system as close as possible to the desirable level and to preserve stability conditions in the presence of unexpected changes in the system behaviour caused by faults. We emphasize that, for this reason, FTC systems have received increased attention in the last decade [3, 27, 39, 41]. The fault-tolerant control approaches can be divided into two main groups, passive and active FTC techniques [5, 39, 47]. Passive FTC techniques, require a priori knowledge about possible (anticipated) faults that can affect the system. They should be taken into consideration in the time of designing such a control system. This leads to unpredictable behaviour in the case of the unanticipated faults. The principal idea of this kind of control is based on treating all possible faults as uncertainties [5, 18]. Mostly, in the case of the passive approach, controller switching techniques are used. In such a case for each possible fault, separate controller exists and the system switch between them to adapt to changes caused by a fault. In turn, active FTC techniques are based on adaptation of the control law using the information given by the fault diagnosis (FD) block (see Fig. 3.14). In such a case it is crucial to design very efficient and robust methods providing the information about the presence of a fault, its localization and size. Such attempts to design FTC systems are currently in the center of interest among researchers and can be easily found in the recent papers, e.g. [25, 26, 44]. The current state-of-the-art in terms of the reconfigurable control design can be illustrated in the form of a graph shown in Fig. 3.15. This is a classification

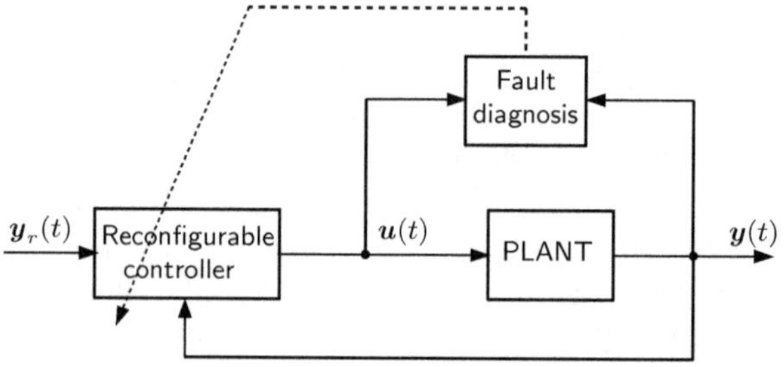

Fig. 3.14 A general scheme of active FTC

Classification of FTC methods
- Mathematical design tools
  - Linear quadratic (LQ)
  - Pseudo-inverse (PI)
  - Intelligent control (IC)
  - Gain scheduling (GS)/Linear parameter varying (LPV)
  - Model following (MF)
  - Adaptive control (AC)
  - Multiple model (MM)
  - Eigenstructure assignment (EA)
  - Feedback linearisation (FL)/dynamic inversion (DI)
  - H_∞ and other robust control techniques
  - Model predictive control (MPC)
  - Quantitative feedback theory (QFT)
  - Linear matrix inequality (LMI)
  - Variable structure control (VSC)/sliding model control (SMC)
  - Generalized internal model control (GIMC)
- Design approaches
  - Pre-computed control laws
    - MM
    - GS/LPV
    - QFT
    - LMI
    - GIMC
  - On-line automatic redesign
    - LQ
    - PI
    - MF/AC
    - EA
    - FL/DI
    - VSC/SMC
    - MPC
- Reconfiguration mechanisms
  - Optimizations
    - LQ
    - H_∞/μ synthesis
    - LMI
    - MPC
  - Switching
    - MM
    - GS/LPV
    - VSC/SMC
  - Matching
    - PI: System matrix
    - EA: Eigenstructure
  - Following
    - MF: State/Output
    - MPC: Set-point/Output
  - Compensation
    - Additive compensation
    - Adaptive compensation
- Types of systems dealt with
  - Linear systems { LQ, PI, MF, EA, MM, MPC, QFT, GIMC
  - Nonlinear systems { GS/LPV, MM, FL/DI, LMI, VSC/SMC, IC

Fig. 3.15 A classification of active fault-tolerant control systems [47]. ©2008 Elsevier. Reproduced with permission of the control design methods currently being developed. However, we should note that in an actual engineering practice, a combination of different methods is often applied at the same time. This allows one to obtain much better results in terms of fault tolerance and to improve the overall performance, stability and fault-tolerance properties of the control system. The methods presented in Fig. 3.15 are described in more detail in the work of Zhang and Jiang [47]. However, it should be mentioned that despite many research efforts and algorithms developed in this field, a majority of the approaches is devoted to the linear systems, whilst the industrial systems are mostly characterized by a nonlinear behaviour. In fact, Fig. 3.15 shows a number of algorithms classified as nonlinear FTC systems. However, linear-parameter-varying approaches or feedback-linearization algorithms are based on approximate models. The method of gain scheduling uses a family of linear controllers. Similarly, a multiple model technique uses a family of possible linear models of a plant. In turn, sliding-mode control requires knowledge of an analytical model of the plant, hence one observes a lack of effective design methods. There are also problems of numerical nature in implementing such algorithms. In other words, there is still a need for developing active fault-tolerant control techniques for nonlinear systems. Indeed, such avenue of research is currently at the forefront of what is happening in FTC, precisely because it is seen as the basis for further growth and development in both theory and practice of the studies into nonlinear control systems and fault identification.

3.4 Conclusions

The chapter provided a brief introduction into the problem of control synthesis describing the background and the most important concepts in terms of the control objectives, and both the open- and the closed-loop realizations. We have also described a general classification of the control strategies. The intention was to emphasize control schemes dedicated to nonlinear systems. With this in mind, a large part of the chapter is devoted to intelligent control designs carried out by means of artificial neural networks. We present the most frequently used schemes: direct control, model reference control, internal model control, feed-forward control, predictive control and optimal control. In each of the presented control strategies, a crucial role is played by a neural network. The examples show that a neural network can be employed in the control scheme not only as the model of the plant but also as the classifier able to map the tracking error into the suitable control surface. We note however that since it is the model-based approaches that are the most popular, the development of a model of the plant — in fact, an inverse model of the plant — is one of the most important tasks to be performed during the control system synthesis. These aspects were discussed in detail in Chap. 2 in Sect. 2.4. Nowadays, both robust and fault-tolerant control are intensively studied and form rapidly developing avenues of research in the modern control theory. The chapter included two sections devoted to robust control and fault-tolerant control, respectively with the purpose of giving a comprehensive review of the described approaches. An analysis of the developed robust control algorithms reveals that most of the approaches are devoted to linear systems. Uncertainty description methods, such as structured and unstructured uncertainty cannot be directly used in the case of nonlinear system representation. Therefore, there is still a need for development of robust control strategies for nonlinear systems. In this context, artificial neural networks and robust model-developing algorithms presented in Chap. 2 in Sect. 2.5 could be made use of. A similar problem is observed in the context of the FTC systems. A great majority of the existing solutions are developed for linear systems. As shown in Fig. 3.15, there are certain approaches developed for nonlinear systems, however we can see that such methods, based on feedback linearization, gains scheduling, application of multi-linear models or linear parameter varying, all use some approximation of the nonlinear model. All the above strongly supports the need for developing intelligent control algorithms, i.e. based on neural networks and modern fault-diagnosis methods.

References

1. Barrett, M.F.: Conservatism with robustness tests for linear feedback control systems. Ph.D. thesis, University of Minnesota (1980)
2. Bemporad, A., Morari, M.: Robust model predictive control: a survey. In: Garulli, A., Tesi, A. (eds.) Robustness in Identification and Control. Lecture Notes in Control and Information Sciences, pp. 207–226. Springer, London (1999)
3. Blanke, M., Kinnaert, M., Lunze, J., Staroswiecki, M.: Diagnosis and Fault-Tolerant Control. Springer, Berlin (2006)
4. Camacho, E.F., Bordóns, C.: Model Predictive Control, 2nd edn. Springer, London (2007)
5. Chen, J., Patton, R.J.: Robust Model-Based Fault Diagnosis for Dynamic Systems. Kluwer, Berlin (1999)
6. Davison, E.J.: The robust control of a servomechanism problem for linear time invariant multivariable systems. IEEE Trans. Autom. Control **21**, 25–34 (1976). First presented at the Alerton Conference (1973)
7. Doyle, J., Glover, K., Khargonekar, P., Francis, B.: State space solutions to standard H_2 and H_∞ control problems. IEEE Trans. Autom. Control **34**, 731–747 (1989)
8. Doyle, J.C.: Guaranteed margins for LQG regulators. IEEE Trans. Autom. Control **23**, 756–757 (1978)
9. Doyle, J.C.: Structured uncertainty in control system design. In: IEE Proceedings, vol. 192-D, pp. 242–250 (1981)
10. Francis, B.A.: A Course in H_∞ Control Theory. Lecture Notes in Control and Information Sciences, vol. 88. Springer, Berlin (1987)
11. Harris, T., Seppala, C., Desborough, L.: A review of performance monitoring and assessment techniques for univariate and multivariate control systems. J. Process Control **9**(1), 1–17 (1999)
12. Hu, H., Liu, J., Wang, L.: Model reference neural network control strategy for flight simulator. In: 2010 IEEE International Conference on Mechatronics and Automation, pp. 1483–1488 (2010)
13. Hunt, K.J., Sbarbaro, D., Zbikowski, R., Gathrop, P.J.: Neural networks for control systems – a survey. Automatica **28**, 1083–1112 (1992)
14. Kalman, R.E.: When is a linear control system optimal. Trans. ASME Ser. D: J. Basic Eng. **86**, 51–60 (1964)

15. Kuschewski, J.G., Hui, S., Żak, S.: Application of feedforward neural network to dynamical system identification and control. IEEE Trans. Neural Netw. **1**, 37–49 (1993)
16. Ławryńczuk, M.: Computationally Efficient Model Predictive Control Algorithms. A Neural Network Approach. Studies in Systems, Decision and Control, vol. 3. Springer, Switzerland (2014)
17. Leigh, J.R.: Control Theory, 2nd edn. IET, London (2008)
18. Liang, Y.: Reliable control of nonlinear systems. IEEE Trans. Autom. Control **45**(4), 706–710 (2000)
19. Lurie, B., Enright, P.: Classical Feedback Control. CRC Press, Boca Raton (2011)
20. Maciejowski, J.: Multivariable Feedback Design. Addison-Wesley, Harlow (1989)
21. Moore, K.: On the history, accomplishments, and future of the iterative learning control paradigm (2009). http://inside.mines.edu/~kmoore/ANNIE.pdf. Presented at the 2009 artificial neural networks in engineering. St Louis Missouri, USA
22. Narendra, K.S., Parthasarathy, K.: Identification and control of dynamical systems using neural networks. IEEE Trans. Neural Netw. **1**, 12–18 (1990)
23. Nise, N.S.: Control Systems Engineering, 6th edn. Wiley, Hoboken (2011)
24. Nørgaard, M., Ravn, O., Poulsen, N., Hansen, L.: Networks for Modelling and Control of Dynamic Systems. Springer, London (2000)
25. Montes de Oca, S., Puig, V., Witczak, M., Dziekan, Ł.: Fault-tolerant control strategy for actuator faults using LPV techniques: application to a two degree of freedom helicopter. Int. J. Appl. Math. Comput. Sci. **22**(1), 161–171 (2012). https://doi.org/10.2478/v10006-012-0012-y
26. Ocampo-Martinez, C., De Dona, J.A., Seron, M.M.: Actuator fault-tolerant control based on set separation. Int. J. Adapt. Control Signal Process. **24**(12), 1070–1090 (2010)
27. Patan, K., Korbicz, J.: Nonlinear model predictive control of a boiler unit: a fault tolerant control study. Appl. Math. Comput. Sci. **22**(1), 225–237 (2012)
28. Pearson, J.B., Staats, P.W.: Robust controllers for linear regulators. IEEE Trans. Autom. Control **19**, 231–234 (1974)
29. Pham, D.T., Xing, L.: Neural Networks for Identification, Prediction and Control. Springer, Berlin (1995)
30. Postlethwaite, I., Edmunds, J.M., MacFarlane, A.G.J.: Principal gains and principal phases in the analysis of linear multivariable feedback systems. IEEE Trans. Autom. Control **26**, 32–46 (1981)
31. Rivals, I., Personnaz, L.: Nonlinear internal model control using neural networks: application to process with delay and design issues. IEEE Trans. Neural Netw. **11**, 80–90 (2000)
32. Rohrs, C.E., Valavani, L., Athans, M.: Convergence studies of adaptive control algorithms: part i. In: Proceedings of the IEEE Conference on Decision and Control, Albuquerque, NM, 10–12 December, pp. 1138–1141 (1980)
33. Rohrs, C.E., Valavani, L., Athans, M., Stein, G.: Robustness of adaptive control algorithms in the presence of unmodeled dynamics. In: Proceedings of the IEEE Conference on Decision and Control, Orlando, FL, 8–10 December, pp. 3–11 (1982)
34. Safonov, M.G.: Robustness and stability aspects of stochastic multivariable feedback system design. Ph.D. thesis, MIT (1977)
35. Safonov, M.G.: Multivariable feedback system design. In: Proceedings of the IEEE Conference on Decision and Control (1981)
36. Safonov, M.G.: Origins of robust control: early history and future speculations. In: Proceedings of the 7th IFAC Symposium on Robust Control Design, Aalborg, Denmark, 20–22 June, pp. 1–8 (2012)
37. Safonov, M.G., Athans, M.: Gain and phase margin for multiloop LQG regulators. IEEE Trans. Autom. Control **22**, 173–178 (1977)
38. Sandberg, I.W.: On the L_2-boundedness of solutions of nonlinear functional equations. Bell Syst. Tech. J. **43**, 1581–1599 (1964)
39. Simani, S., Farsoni, S.: Fault Diagnosis and Sustainable Control of Wind Turbines. Robust Data-Driven and Model-Based Strategies. Butterworth-Heinemann, Oxford (2018)

40. Skogestad, S., Postlethwaite, I.: Multivariable Feedback Control. Analysis and Design, 2nd edn. Wiley, New York (2005)
41. Staroswiecki, M., Yang, H., Jiang, B.: Active fault tolerant control based on progressive accomodation. Automatica **43**(12), 2070–2076 (2007)
42. Tsai, M.C., Gu, D.W.: Robust and Optimal Control. A Two-Port Framework Approach. Advances in Industrial Control. Springer, London (2014)
43. Wong, P.K.: On the interaction structure of linear multi-input feedback control systems. Master's thesis, MIT (1975)
44. Yetendje, A., Seron, M., De Doná, J.: Robust multisensor fault tolerant model-following MPC design for constrained systems. Int. J. Appl. Math. Comput. Sci. **22**(1), 211–223 (2012). https://doi.org/10.2478/v10006-012-0016-7
45. Zames, G.: Functional analysis applied to nonlinear feedback systems. IEEE Trans. Circuit Theory **CT-10**, 392–404 (1963)
46. Zames, G.: Feedback and optimal sensitivity: model reference transformations, multiplicative seminorms, and approximate inverses. IEEE Trans. Autom. Control **26**, 301–320 (1981)
47. Zhang, Y., Jiang, J.: Bibliographical review on reconfigurable fault-tolerant control systems. Annu. Rev. Control **32**(2), 229–252 (2008)

Chapter 4
Model Predictive Control

4.1 Introduction

Model predictive control has been the subject of intensive research for the last three decades [8, 25, 47]. This culminated in many practical applications [1, 7, 20]. The attractiveness of predictive control algorithms comes from their ability to consider process and technological constraints imposed on input, output or state variables. Another very important reason is that the operating principles are understandable and relatively easy to explain to practitioners, which seems to be a crucial aspect when it comes to an industrial implementation of a new control scheme. This explains the wide-spread use of predictive control algorithms in the petro-chemical and related industries, where the fact that the imposed constraints are made sure to be satisfied is particularly important.

Predictive control is a modern control strategy that derives the control sequence by solving at each sampling time a finite-horizon open-loop optimal control problem. The important role is played by a model of the process used to predict future plant outputs based on the past and current outputs as well as on future control signals. These are calculated by means of the minimization of the cost function taking into account constraints. In many cases, the constraints are in the form of inequalities imposed on the process variables, e.g. the input saturations. If such nonlinearities are not taken into account, it can result in a degraded performance of closed-loop control and lead to stability problems. As pointed out in the outstanding survey papers of Morari and Lee [27] and Mayne and colleagues [26], the predictive control theory has reached a high level of maturity, especially in the case of linear systems. However, nonlinear systems remain problematic, for a number of reasons. The most common issues arise in relation to the following areas: the modelling of nonlinear processes, state estimation, robustness or fault-tolerant control [27].

The ideas described in this chapter can be viewed as arising as a response to the emerging problems encountered in predictive control. In order to deal with nonlinearities, artificial neural networks are used. The neural model can be used as a one-step ahead predictor, which is then run recursively to obtain a k-step ahead

© Springer Nature Switzerland AG 2019

K. Patan, *Robust and Fault-Tolerant Control*, Studies in Systems, Decision and Control 197, https://doi.org/10.1007/978-3-030-11869-3_4

prediction of the plant output. As discussed in Chap. 2, the neural network models are well-suited to play such a role. We also investigate the problem of the stability of a predictive scheme based on the input-output model and show that a cost function is monotonically decreasing with respect to time. The derived stability conditions can be further applied to redefine a constrained optimization problem in order to calculate a control sequence. The fault tolerance of a control system constitutes yet another important problem area. We note that MPC seems to be very suitable for FTC: some faults can be represented through a modification of constraints in the MPC algorithm, while others can be handled through a modification of the system model. Except for certain robust conservative solutions, MPC, despite its sophisticated operation idea and advantages, is not able to handle sensor or actuator faults. The chapter proposes a methodology of compensation for sensor faults by introducing fault detection capabilities as well as supervisory units into the control system. The last problem discussed in the chapter concerns the robustness of MPC. Generally, model uncertainty has two main sources: unmodelled dynamics of a plant and unmeasured noise/disturbances entering the plant. The chapter presents certain results related to the area of robust control based on deriving uncertainty associated with the model.

4.2 Nonlinear Model Predictive Control

In MPC, a control sequence is calculated by solving on-line a finite horizon optimal control problem subject to constraints. The derived control is applied to the plant and the cycle repeats. Here, the cost function to be minimized is defined as follows:

$$J = \sum_{i=1}^{N_p} e^2(k+i) + \rho \sum_{i=1}^{N_u} \Delta u^2(k+i-1), \tag{4.1}$$

where $e(k+i) = y_r(k+i) - \hat{y}(k+i)$ is the tracking error, $y_r(k+i)$ is the future reference signal, $\hat{y}(k+i)$ is the prediction of future plant outputs, $\Delta u(k+i-1) = u(k+i-1) - u(k+i-2)$, N_p is the prediction horizon, N_u is the control horizon ($N_u < N_p$) and ρ represents a factor penalizing changes in the control signal. In the vector form, the cost (4.1) can be rewritten as:

$$J = (\boldsymbol{y}_r - \boldsymbol{y})^\mathsf{T}(\boldsymbol{y}_r - \boldsymbol{y}) + \rho \Delta \boldsymbol{u}^\mathsf{T} \Delta \boldsymbol{u}, \tag{4.2}$$

where $\boldsymbol{y} = [\hat{y}(k+1), \ldots, \hat{y}(k+N_p)]^\mathsf{T}$ is the vector of predictions of the plant output, $\boldsymbol{y}_r = [y_r(k+1), \ldots, y_r(k+N_p)]^\mathsf{T}$ is the future reference trajectory, and $\Delta \boldsymbol{u} = [\Delta u(k), \ldots, \Delta u(k+N_u-1)]^\mathsf{T}$ represents future control moves. Our goal is to minimize, at each sampling time, the criterion (4.1) with respect to N_u future controls,

$$\boldsymbol{u}(k) = [u(k), \ldots, u(k+N_u-1)]^\mathsf{T}. \tag{4.3}$$

We emphasize that the appealing property of MPC is its ability to consider various constraints. Usually, it is assumed that beyond the control horizon the control remains constant. This constitutes the following constraints:

$$\Delta u(k+i) = 0, \quad i \geq N_u. \tag{4.4}$$

In industrial installations, constraints related to the control signal are usually related to actuator properties, thus

$$u_{min} \leq u(k+i) \leq u_{max}, \quad i = 0, \ldots, N_u - 1, \tag{4.5}$$

where u_{min} and u_{max} represent the minimal and the maximal values of the control signal, respectively. In practical applications, the output of the system is also bounded, thus we have:

$$y_{min} \leq \tilde{y}(k+i) \leq y_{max}, \quad i = 1, \ldots, N_y, \tag{4.6}$$

where y_{min} and y_{max} are the minimal and the maximal values of the plant output, respectively and N_y is the output constraint horizon. However, from the point of view of implementation, the output constraints are often regarded as so-called soft constraints [49]:

$$y_{min} - \epsilon_{min}(k+i) \leq \tilde{y}(k+i) \leq y_{max} + \epsilon_{max}(k+i), \quad i = 1, \ldots, N_y, \tag{4.7}$$

where $\epsilon_{min}(k+i)$ and $\epsilon_{max}(k+i)$ are slack variables.

For nonlinear systems, the optimization problem (4.1) with the constraints (4.4)–(4.7) has to be solved at each sample time, giving a sequence of future controls $\boldsymbol{u}(k)$, where the first element is taken to control the plant. One of the distinguishing characteristics of MPC is the presence of what is known as the receding horizon: the control signal is derived in such a way as to achieve the desired behaviour of the control system in the subsequent N_p time steps. Yet another important property is the so-called control horizon $N_u < N_p$, meaning that only the first N_u future controls are determined. From that point onwards, the control is assumed to remain constant.

For nonlinear systems, the optimization of the cost (4.1) subject to the constraints (4.4), (4.5) and (4.6) (or (4.7)) constitutes a nonlinear programming task, and finding a solution for such a task can be troublesome. For the most part, the issues are related to the following: (i) the multimodality of the cost, (ii) the computational burden of a nonlinear optimization procedure, (iii) problems with the convergence of a procedure to an acceptable solution. Furthermore, the fundamental issue in model predictive control is to design an accurate model of the controlled process. All these topics are discussed in the following sections.

4.2.1 System Modelling

Let us consider a nonlinear system represented as:

$$y(k+1) = f(\varphi(k)), \tag{4.8}$$

where f is a nonlinear function, $\varphi(k)$ is a regression vector of the form $\varphi(k) = [y(k), \ldots, y(k-n_a+1), u(k), \ldots, u(k-n_b+1)]^{\mathsf{T}}$, n_a and n_b represent the number of past outputs and inputs needed for designing the model, respectively. It should be kept in mind that the precise values of f, $y(k)$ and $\varphi(k)$ are in general not known, however a set of noise-corrupted measurements $\tilde{y}(k)$ and $\tilde{\varphi}(k)$ of $y(k)$ and $\varphi(k)$, respectively, is available. We have:

$$\tilde{y}(k+1) = f(\tilde{\varphi}(k)) + \varepsilon(k), \tag{4.9}$$

where $\varepsilon(k)$ stands for the additive white measurement noise. Our goal is to find an estimate $\hat{f}$ of the nonlinear function f in such a way as to achieve possibly minimal modelling error $|f - \hat{f}|$. This estimate is usually obtained by a prediction error method. Among many approaches reported in the literature, it is also artificial neural networks that can be successfully applied in this context [14, 28]. In the field of neural modelling, the simplest solution is to use feed-forward networks with external dynamics [29, 33] (see Chap. 2, Sect. 2.3.1). There are two strong arguments for using this structure. First, such a topology is able to approximate any nonlinear mapping with arbitrary accuracy [16]. The second argument is that such a structure is the most popular within the control community as it is very simple to both train and implement. In such a situation a neural model with one hidden layer is considered:

$$\hat{y}(k+1) = \hat{f}(\tilde{\varphi}(k)) = \sigma_o(\boldsymbol{W}_2^{\mathsf{T}}\boldsymbol{\sigma}_h(\boldsymbol{W}_1^{\mathsf{T}}\tilde{\varphi}(k) + \boldsymbol{b}_1) + \boldsymbol{b}_2), \tag{4.10}$$

where $\boldsymbol{W}_1 \in \mathbb{R}^{(n_a+n_b)\times v}$ and $\boldsymbol{W}_2 \in \mathbb{R}^{v\times 1}$ are weight matrices of the hidden and output layers, respectively, $\boldsymbol{b}_1 \in \mathbb{R}^v$ and $\boldsymbol{b}_2 \in \mathbb{R}^1$ are bias vectors of the hidden and output units, respectively, $\boldsymbol{\sigma}_h : \mathbb{R}^v \to \mathbb{R}^v$ is the vector-valued activation function of the hidden layer, and $\sigma_o : \mathbb{R}^1 \to \mathbb{R}^1$ is the activation function of the output layer and v stands for the number of hidden neurons. Usually, training of the neural model is carried out using data recorded in the plant working at normal operating conditions either in the open- or closed-loop control depending on the exact conditions.

When a disturbance model is also considered, the entire model is represented as:

$$\hat{y}(k+1) = \hat{f}(\tilde{\varphi}(k)) + d(k), \tag{4.11}$$

where $d(k)$ stands for the disturbance model defined as:

$$d(k) = k_c\left(\tilde{y}(k) - \hat{y}(k)\right) - d(k-1), \tag{4.12}$$

where k_c represents the gain of the disturbance model. As the disturbance model (4.12) includes an integrator, offset-free steady-state behaviour of the control system can be achieved [8, 48]. It is often the case that k_c is assumed to be equal to 1 and $d(k)$ is assumed to be constant within the prediction horizon [8].

The i-step ahead predictor ($i = 1, \ldots, N_p$) can be obtained in several ways:

1. by successive recursion of a one-step ahead nonlinear model,
2. by designing N_p predictors, each providing prediction of the plant output at the suitable time instant,
3. by instantaneous linearization of a nonlinear model of a plant.

In the first case, the control system uses a nonlinear model of a plant, thus the minimization of the objective function has to be carried out through an iterative procedure. Unfortunately, successive recursion means that the prediction error associated to the one-step ahead predictor rapidly increases when the i-step ahead predictor is in use. In the second case, we no longer encounter problems with a recursively increasing prediction error but the design process is a time-consuming and difficult task as it requires one to develop N_p predictors. In the third case, the unique solution of an optimization problem exists and a future control signal can be calculated directly. Unfortunately, the instantaneous linearization technique suffers from some drawbacks. It relies on the linearized model, which may have a limited validity in certain operating ranges. In this section, the i-step ahead prediction of the process output is calculated using successive recursion of (4.10) according to the formula:

$$\hat{y}(k+i) = \hat{f}(\tilde{\varphi}(k+i-1)) \quad \text{for } i \in [1, N_p]. \tag{4.13}$$

It should be pointed out that measurements of the process output are available up to time k. Therefore, one should substitute predictions for actual measurements, since these do not exist:

$$y(k+i) = \hat{y}(k+i), \quad \forall i > 1. \tag{4.14}$$

In order to realize the i-step ahead predictor, one needs to feed the network with the past predicted outputs according to (4.13) and (4.14), and a feed-forward network trained using a series-parallel identification model can turn out to be insufficient for this task. The solution is to use the parallel identification model or a recurrent neural network, e.g. a neural network output error (NNOE) model represented by (4.10) but with the regressor:

$$\tilde{\varphi}(k) = [\hat{y}(k), \ldots, \hat{y}(k-n_a+1), u(k), \ldots, u(k-n_b+1)]^{\mathsf{T}}. \tag{4.15}$$

4.2.2 Uncertainty Handling

A control system is called robust, if it is insensitive to the differences observed between the plant and the model used for the controller synthesis. Every modelling procedure, either for linear or nonlinear processes suffers, from the so-called model mismatch, i.e. the case that the model of the system is not a faithful replica of plant dynamics. On this basis, uncertainty can be seen as a measure of the unmodelled dynamics, noise and disturbances affecting the plant. Moreover, it should be noted that if a model is derived in closed-loop control (with a preliminary selected controller), its quality is limited to the properties of the controller used to record the data. When such a model is used to design MPC, the quality of control can be called into question. Consequently, a model should be improved in one way or the other. In this section, we propose a possible way of identifying uncertainty related to neural models.

Let us consider the following uncertain nonlinear and discrete-time system :

$$y(k) = f(\varphi(k)) + w(k), \tag{4.16}$$

where $w(k)$ represents the uncertainty, assumed to be additive and bounded by a compact set $\mathcal{W}$ ($w(k) \in \mathcal{W}$). This kind of uncertainties are often called global [8]. At first glance, global uncertainty seems to be more like a disturbance than uncertainty, since it influences the system similarly to external perturbations. However, $w(t)$ can be expressed as a function of the past system inputs and outputs and (4.16) rewritten as follows:

$$y(k) = f(\varphi(k)) + \Delta f(\varphi(k)) = \bar{f}(\varphi(k)). \tag{4.17}$$

Additive uncertainty can represent a wide class of model mismatches [24].

Starting from (4.16) and using the nominal model of the plant together with the available measurements, uncertainty can be estimated as:

$$w(k) = y(k) - f(\varphi(k)) \approx \tilde{y}(k) - \hat{f}(\varphi(k)) = \tilde{y}(k) - \hat{y}(k). \tag{4.18}$$

The equation (4.18) looks like a residual, that is a signal defined as a difference between the measured process output and the output of a model (frequently used for fault-diagnosis purposes). From our perspective however, there is one distinct characteristic that differentiates an uncertainty estimate from a residual — the uncertainty estimate represents the dynamics of the changes occurring in the plant as the effect of structural or parametric uncertainty. Based on this estimate, it is possible to predict uncertainty in the future, which is required for the control synthesis. The residual, on the other hand, is calculated on-line at each sampling time.

In order to represent the widest possible class of model mismatches, the modification of the so-called model error modelling introduced in Chap. 2, Sect. 2.5.2 can be used. The entire procedure can be viewed as a robust identification method of the passive kind. A method uses the nominal model of a plant, which is developed without uncertainty considerations (as discussed e.g. in Sect. 4.6.1). Then, the uncertainty model is designed by analyzing the signal $w(k)$ evaluated from the inputs [32].

In general, the error model can be described by the following difference equation:

$$\hat{w}(k+1) = g(\phi(k)), \tag{4.19}$$

where $\hat{w}(k+1)$ represents an uncertainty estimate at the time instant $k+1$, $\phi(k) = [w(k), \ldots, w(k-n_c+1), u(k), \ldots, u(k-n_d+1)]^\mathsf{T}$, n_c and n_d represent the number of past uncertainty estimates and inputs needed for designing the error model, respectively. Applying a neural network with external dynamics to represent a function $g(\cdot)$, the following error model is achieved:

$$\hat{w}(k+1) = \sigma_o(\boldsymbol{V}_2^\mathsf{T}\boldsymbol{\sigma}_h(\boldsymbol{V}_1^\mathsf{T}\phi(k) + \boldsymbol{\beta}_1) + \boldsymbol{\beta}_2), \tag{4.20}$$

where $\boldsymbol{V}_1 \in \mathbb{R}^{(n_c+n_d)\times v_w}$ and $\boldsymbol{V}_2 \in \mathbb{R}^{v_w\times 1}$ are weight matrices of the hidden and output layers, respectively, $\boldsymbol{\beta}_1 \in \mathbb{R}^{v_w}$ and $\boldsymbol{\beta}_2 \in \mathbb{R}^1$ are bias vectors of the hidden and output units, respectively, $\boldsymbol{\sigma}_h : \mathbb{R}^{v_w} \rightarrow \mathbb{R}^{v_w}$ is the activation function of the hidden layer, and $\sigma_o : \mathbb{R}^1 \rightarrow \mathbb{R}^1$ is the activation function of the output layer and v_w stands for the number of hidden neurons of the error model. According to (4.16), the final representation of a robust model is given by:

$$\bar{y}(k) = \hat{y}(k) + \hat{w}(k). \tag{4.21}$$

As we cannot be sure that the error model output satisfies the normal distribution property, to determine the confidence region associated with the model the rule-of-thumb can be applied. This leads to the three-sigma algorithm giving approximately 99.7% of the confidence interval. The confidence region forms uncertainty bounds around the response of the robust model: the upper band

$$\overline{w}(k) = \bar{y}(k) + 3\sigma \tag{4.22}$$

and the lower band

$$\underline{w}(k) = \bar{y}(k) - 3\sigma, \tag{4.23}$$

where σ is the standard deviation of $\hat{w}(k)$ calculated over a testing set of data. The centre of the uncertainty region is $\bar{y}(k)$ [32]. As the *error model* is the estimate of the fundamental model mismatches, we should prepare a suitable training data. This problem is exhaustively discussed in Sect. 4.6.

4.2.3 Stability Analysis

In general, the stability of control systems is a problem of crucial importance in control theory. The literature on stability of model predictive control of dynamic systems, linear and nonlinear, is vast. The exhaustive review of the solutions for achieving stability can be found by the interested reader in the work of Mayne and colleagues [26]. Arguably, the most popular techniques reported in the literature can be divided into two main classes. The first class contains techniques using a cost function as a Lyapunov candidate function. This includes the following: terminal constraint [18], infinite output prediction horizon [18], terminal cost function [41], and terminal constraint set methods [43]. The second class requires that the state is decreasing in some norm, e.g. $\|\cdot\|_1$, or $\|\cdot\|_\infty$ [3].

Owing to certain interesting properties of state-space approaches, the majority of the articles in this field is devoted to systems represented in the state-space. As a consequence, the cost function is therefore based on the state. Since we decided to investigate a methodology based on GPC, the solutions proposed for the systems represented in the state-space cannot be applied. The problem of the stability of GPC was successfully addressed in a number of publications, where a number of potential solutions was proposed, e.g. infinite-horizon GPC (GPC$^\infty$) [42], constrained receding-horizon predictive control (CRHPC) [9], stable generalized predictive control (SGPC) [13] or min-max GPC [19]. Note that these articles dealt with the problem of stability of linear input/output systems. The stability is achieved by imposing terminal constraints on the inputs and outputs over some constraint horizon. Since in this chapter, we are interested in predictive control of a nonlinear process using nonlinear predictor based on a dynamic neural network, stability is investigated by checking the monotonicity of the cost. This approach is similar as in the case of the method employed in the paper by Scokaert and Clark [42] but in our case the predictor is nonlinear, the prediction horizon is of a finite value, and the control horizon is not greater than the prediction horizon.

Let us consider the nonlinear model predictive control based on the following open-loop optimization problem:

$$
\begin{aligned}
& \boldsymbol{u}(k) \triangleq \arg\min_{\boldsymbol{u}} J, && (4.24a)\\
\text{s.t.}\quad & e(k+N_p+j)=0, \quad \forall j \in [1, N_c], && (4.24b)\\
& \Delta u(k+N_u+j)=0, \quad \forall j \geq 0, && (4.24c)\\
& u_{min} \leq u(k+j) \leq u_{max}, \quad \forall j \in [0, N_u-1], && (4.24d)
\end{aligned}
$$

where N_c is the constraints horizon, u_{min} and u_{max} are lower and upper control bounds.

Theorem 4.1 ([34]) *The nonlinear model predictive control system (4.24) using a predictor based on (4.13) is asymptotically stable, if the following conditions are satisfied:*

(i) $\rho \neq 0$,
(ii) $N_c = \max\left[n_a+1, n_b+1+N_u-N_p\right]$,

regardless the choice of N_p and N_u.

Proof The cost function at time k has the following form:

$$
J(k) = \sum_{i=1}^{N_p} e^2(k+i) + \rho \sum_{i=1}^{N_u} \Delta u^2(k+i-1). \tag{4.25}
$$

Let us assume that $\boldsymbol{u}(k)$ is the optimal control at time k found by an optimization procedure. Now, let us introduce the suboptimal control $\boldsymbol{u}^*(k+1)$ postulated at time $k+1$:

$$\boldsymbol{u}^*(k+1) = [u(k+1), \ldots, u(k+N_u-1), u(k+N_u-1)]^\mathsf{T}.$$

The control sequence $\boldsymbol{u}^*(k+1)$ is formed based on the control derived at time k therefore, assuming that unmeasured disturbances $d(k)$ are constant within the prediction horizon, the predictions $y(k+i)$ derived at the time $k+1$ are the same as these derived at the time k. Therefore, for the suboptimal control $\boldsymbol{u}^*(k+1)$, the cost function can be defined as follows:

$$J^*(k+1) = \sum_{i=2}^{N_p+1} e^2(k+i) + \rho \sum_{i=2}^{N_u} \Delta u^2(k+i-1). \tag{4.26}$$

The difference of cost functions $J(k)$ and $J^*(k+1)$ can be defined as:

$$J^*(k+1) - J(k) = e^2(k+N_p+1) - e^2(k+1) - \rho \Delta u^2(k). \tag{4.27}$$

Taking into account a set of terminal equality constraints (4.24b), it is obvious that $e(k+N_p+1) = 0$, then

$$J^*(k+1) - J(k) = -e^2(k+1) - \rho \Delta u^2(k) \leq 0. \tag{4.28}$$

Analyzing the prediction equation for time instants beyond the prediction horizon:

$$\begin{aligned}\hat{y}(k+N_p+j) = f(&y(k+N_p+j-1), \ldots, y(k+N_p+j-n_a),\\ &u(k+N_p+j-1), \ldots, u(k+N_p+j-n_b)) + d(k),\end{aligned} \tag{4.29}$$

one can conclude that tracking error equality constraints hold for all $j \geq 1$, if:

(i) $N_c \geq n_a + 1$, assuming that $n_a \geq n_b + 1 + N_u - N_p$.
(ii) $N_c \geq n_b + 1 + N_u - N_p$, assuming that $n_a < n_b + 1 + N_u - N_p$.

Using both results, setting the constraint horizon on the value

$$N_c = \max\left[n_a + 1, n_b + 1 + N_u - N_p\right]$$

guarantees that tracking error equality constraints hold not only for $j \in [1, N_c]$ but for all $j \geq 1$. Moreover, looking at the definition of $\boldsymbol{u}^*(k+1)$, inequality constraints (4.24d) at the time $k+1$ are also satisfied. Using these considerations one can state that $\boldsymbol{u}^*(k+1)$ satisfies all constraints at time $k+1$, and subsequently the vector $\Delta\boldsymbol{u}^*(k+1)$ also satisfies constraints (4.24c). Furthermore, if $\boldsymbol{u}(k+1)$ is the optimal solution of the optimization problem time $k+1$, then $J(k+1) \leq J^*(k+1)$ as $\boldsymbol{u}^*(k+1)$ is the suboptimal one. Finally

$$\Delta J(k+1) \leq -e^2(k+1) - \rho \Delta u^2(k) \tag{4.30}$$

and it is clear that for $\rho \neq 0$ the cost is monotonically decreased with respect to time and the control system is stable.

Remark 4.2 The presented results were achieved with the assumption that a set of feasible solutions of the optimization problem exists. As the considered model predictive control requires one to solve a nonlinear optimization problem with constraints, the problem of feasibility is non-trivial. Note that the feasibility considerations are not a focal point of our investigations, however some numerical algorithms for solving the optimization problem are proposed and discussed later on in this section.

4.2.4 Nonlinear Optimization with Constraints

When a nonlinear neural network is applied as a process model, the output predictions are nonlinear in the control inputs. This constitutes a complex nonlinear programming problem, which should be solved in real-time while the optimization procedure should assure fast convergence and numerical robustness. The numerical algorithm suitable for solving the problem should be characterized by fast convergence, real-time processing and numerical robustness. Therefore, second-order optimization algorithms are a reasonable choice. Here, a combination of the Newton and Levenberg–Marquardt methods are used as proposed in [29, 34, 36, 37], which seems to be a reasonable choice. The update rule can be represented as:

$$\boldsymbol{u}^{(i+1)} = \boldsymbol{u}^{(i)} - \left(\boldsymbol{H}^{(i)} + \lambda^{(i)}\mathbf{I}\right)^{-1} \boldsymbol{G}^{(i)}, \tag{4.31}$$

where $\boldsymbol{u}^{(i)}$ is the current iteration of the sequence of future control inputs, $\boldsymbol{H}^{(i)}$ is a Hessian calculated at an ith iteration, $\boldsymbol{G}^{(i)}$ is the gradient vector derived at an ith iteration, $\lambda^{(i)}$ is the parameter used to ensure the positive definiteness of the Hessian matrix, and $\mathbf{I}$ stands for the identity matrix.

Let us rewrite the optimization problem (4.24) as follows:

$$\boldsymbol{u}(k) \triangleq \arg\min_{\boldsymbol{u}} J, \tag{4.32a}$$

$$\text{s.t.}\quad \boldsymbol{h}(\boldsymbol{u}) = 0, \tag{4.32b}$$

$$\boldsymbol{g}(\boldsymbol{u}) \leq 0, \tag{4.32c}$$

where elements of $\boldsymbol{h}(\boldsymbol{u})$ have the form:

$$h_i = \begin{cases} e(k+N_p+i), & \text{for } i = 1, \ldots, N_c \\ \Delta u(k+N_u+i), & \text{for } i > N_c \end{cases},$$

and elements of $\boldsymbol{g}(\boldsymbol{u})$ are defined as:

$$g_i = \begin{cases} u(k+i) - u_{max}, & \text{for } i = 0, \ldots, N_u - 1 \\ u_{min} - u(k+i), & \text{for } i = N_u, \ldots 2N_u - 1 \end{cases}.$$

Suppose $\boldsymbol{u}_0$ is a solution of the problem (4.32), and $A(\boldsymbol{u})$ is a set of active inequality constraints. Assuming gradients of $\boldsymbol{h}$ and $\boldsymbol{g}_i, i \in A(\boldsymbol{u})$ evaluated at $\boldsymbol{u}_0$ are linearly independent, $\boldsymbol{u}_0$ is said to be regular, and we assume also that there exist Lagrange multiplier $\boldsymbol{\mu} \geq 0$ and $\boldsymbol{\lambda} \geq 0$, such that the Lagrangian

$$L(\boldsymbol{u}, \boldsymbol{\mu}, \boldsymbol{\lambda}) = J(\boldsymbol{u}) + \boldsymbol{\mu}^T \boldsymbol{h}(\boldsymbol{u}) + \boldsymbol{\lambda}^T \boldsymbol{g}(\boldsymbol{u}) \tag{4.33}$$

is stationary with respect to $\boldsymbol{u}$ at $\boldsymbol{u}_0$. Based on the Lagrangian (4.33), the problem (4.32) can be solved using the first-order Karush-Kuhn-Tucker (KKT) conditions [5]. Unfortunately, a direct analytical solution of the problem resulting from KKT conditions is extremely hard or even impossible to derive. The most popular methods for solving nonlinear programming problems are gradient projection and various penalty function methods. In general, gradient projection methods possess the disadvantage that feasibility needs to be maintained during optimization. From the numerical-computation point of view, after a trial step, the solution has to be altered slightly to maintain feasibility. On the one hand, a derivation of the projection operator in the case of nonlinear constraints is a challenging problem. On the other hand, penalty-function methods make it possible to transform the original nonlinear programming problem into an unconstrained one, which can be solved using classical algorithms.

4.2.5 Terminal Constraints Handling

A popular approach for considering equality constraints is to transform the original problem to its alternative, unconstrained, form. Substitute

$$\bar{J} = J + \mu \sum_{i=1}^{N_c} h_i^2, \tag{4.34}$$

then, the objective is to solve the unconstrained problem:

$$\boldsymbol{u}(k) \triangleq \arg\min_{\boldsymbol{u}} \bar{J}, \tag{4.35}$$

where μ is a suitably large constant. At this stage, let us assume that the inequality constraints (4.32c) are not taken into account. If $\boldsymbol{u}_\mu$ is a solution of the problem (4.33), it can be shown [51] that as $\mu \to \infty$ there obtains $\boldsymbol{u}_\mu \to \boldsymbol{u}^*$, where $\boldsymbol{u}^*$ is a solution of (4.32). Now, one can use the Newton-based algorithm introduced in Sect. 4.2 to minimize the cost (4.34). Firstly, the elements of a gradient $\boldsymbol{G}$ are represented by the formula:

$$G_j = \frac{\partial \bar{J}}{\partial u(k+j)} = \sum_{i=N_1}^{N_2} \frac{\partial e^2(k+i)}{\partial u(k+j)} + \rho \sum_{i=1}^{N_u-1} \frac{\partial \Delta u^2(k+i-1)}{\partial u(k+j)} + \mu \sum_{i=1}^{N_c} \frac{\partial e^2(k+N_2+i)}{\partial u(k+j)}, \tag{4.36}$$

where $j = 0, \ldots, N_u - 1$ and partial derivatives are calculated as follows:

- the partial derivative $\frac{\partial e^2(k+i)}{\partial u(k+j)}$

$$\frac{\partial e^2(k+i)}{\partial u(k+j)} = 2e(k+i)\frac{\partial e(k+i)}{\partial u(k+j)} = -2e(k+i)\frac{\partial \hat{y}(k+i)}{\partial u(k+j)}, \tag{4.37}$$

- the partial derivative $\frac{\partial \Delta u^2(k+i-1)}{\partial u(k+j)}$

$$\frac{\partial \Delta u^2(k+i-1)}{\partial u(k+j)} = 2\Delta u(k+i-1)\frac{\partial \Delta u(k+i-1)}{\partial u(k+j)}, \tag{4.38}$$

where

$$\frac{\partial \Delta u(k+i-1)}{\partial u(k+j)} = \begin{cases} 1 & \text{for } j = i \\ -1 & \text{for } j = i-1 \end{cases}, \tag{4.39}$$

- the partial derivative $\frac{\partial e^2(k+N_2+i)}{\partial u(k+j)}$ can be derived in a manner analogous to (4.37).

In turn, the elements of the Hessian $\boldsymbol{H}$ have the form:

$$H_{jp} = \frac{\partial^2 \bar{J}}{\partial u(k+j)\partial u(k+p)} = \sum_{i=N_1}^{N_2} \frac{\partial^2 e^2(k+i)}{\partial u(k+j)\partial u(k+p)} + \rho \sum_{i=1}^{N_u-1} \frac{\partial^2 \Delta u^2(k+i-1)}{\partial u(k+j)\partial u(k+p)} + \mu \sum_{i=1}^{N_c} \frac{\partial^2 e^2(k+N_2+i)}{\partial u(k+j)\partial u(k+p)}, \tag{4.40}$$

where $j = 0, \ldots, N_u - 1$, $p = 0, \ldots, N_u - 1$ and partial derivatives are calculated as follows:

- the partial derivative $\frac{\partial^2 e^2(k+i)}{\partial u(k+j)\partial u(k+p)}$

$$\frac{\partial^2 e^2(k+i)}{\partial u(k+j)\partial u(k+p)} = 2\frac{\partial \hat{y}(k+i)}{\partial u(k+j)}\frac{\partial \hat{y}(k+i)}{\partial u(k+p)} - 2e(k+i)\frac{\partial^2 \hat{y}(k+i)}{\partial u(k+j)\partial u(k+p)}, \tag{4.41}$$

- the partial derivative $\frac{\partial^2 \Delta u^2(k+i-1)}{\partial u(k+j)\partial u(k+p)}$

$$\frac{\partial^2 \Delta u^2(k+i-1)}{\partial u(k+j)\partial u(k+p)} = 2\frac{\partial \Delta u(k+i-1)}{\partial u(k+j)}\frac{\partial \Delta u(k+i-1)}{\partial u(k+p)}, \tag{4.42}$$

- the partial derivative $\frac{\partial^2 e^2(k+N_2+i)}{\partial u(k+j)\partial u(k+p)}$ is calculated analogously to (4.41).

To complete the calculations, it is required to gather the partial derivatives: $\frac{\partial \hat{y}(k+i)}{\partial u(k+j)}$, $\frac{\partial \hat{y}(k+i)}{\partial u(k+p)}$, and $\frac{\partial^2 \hat{y}(k+i)}{\partial u(k+j)\partial u(k+p)}$. These derivatives can be easily calculated based on neural predictor (4.13) and the fact that derivatives of the commonly used activation functions can be easily derived. However, for want of space, these formulae are not described here. They can be found in [29], pages 183–186.

Algorithm 4.1 Solution projection

```
for i := 0 to N_u − 1 do
    if u(k + i) > u_max then
        | u(k + i) := u_max
    else if u(k + i) < u_min then
        | u(k + i) := u_min
    end
end
```

An another problem is to find a way of taking into account the inequality constraints (4.32c). The simplest answer is to project the solution into a feasible region as proposed in Algorithm 4.1. This simple prescription can [deteriorate the optimal solution] achieved by minimizing the criterion (4.34) but guarantees that the inequality constraints are satisfied.

4.2.6 A Complete Optimization Procedure

Let us introduce the following penalty cost function:

$$\tilde{J} = \bar{J} + \lambda \sum_{i=0}^{2N_u-1} g_i^2(\boldsymbol{u})S(g_i(\boldsymbol{u})), \tag{4.43}$$

where

$$S(x) = \begin{cases} 1 & \text{if } x > 0, \\ 0 & \text{if } x \leq 0. \end{cases} \tag{4.44}$$

The function $S(x)$ makes it possible to consider a set of active constraints $A(\boldsymbol{u})$ at a current iteration of the algorithm. Thus, in contrast to the algorithm presented in the previous section, all the active inequality constraints are taken into account during the optimization phase. The objective is to solve the unconstrained problem:

$$\boldsymbol{u}(k) \triangleq \arg\min_{\boldsymbol{u}} \tilde{J}. \tag{4.45}$$

In the considered case, the elements of the gradient $\boldsymbol{G}$ are represented by the formula:

$$G_j = \frac{\partial \tilde{J}}{\partial u(k+j)} = \frac{\partial \bar{J}}{\partial u(k+j)} + 2\lambda \sum_{i=0}^{2N_u-1} g_i(\boldsymbol{u}) S(g_i(\boldsymbol{u})) \frac{\partial g_i(\boldsymbol{u})}{\partial u(k+j)}, \tag{4.46}$$

where $j = 0, \ldots, N_u - 1$. The first partial derivative in (4.46) can be derived using (4.36), and the partial derivatives $\frac{\partial g_i(\boldsymbol{u})}{\partial u(k+j)}$ are defined as follows:

$$\frac{\partial g_i(\boldsymbol{u})}{\partial u(k+j)} = \begin{cases} 1 & \text{for } i = j, i = 0, \ldots, N_u - 1 \\ -1 & \text{for } i = j, i = N_u, \ldots, 2N_u - 1 \\ 0 & \text{for } i \neq j \end{cases}. \tag{4.47}$$

In turn, the elements of the Hessian $\boldsymbol{H}$ have the form:

$$\begin{aligned} H_{jp} =& \frac{\partial^2 \tilde{J}}{\partial u(k+j)\partial u(k+p)} = \frac{\partial^2 \bar{J}}{\partial u(k+j)\partial u(k+p)} \\ &+ 2\lambda \sum_{i=0}^{2N_u-1} S(g_i(\boldsymbol{u})) \frac{\partial g_i(\boldsymbol{u})}{\partial u(k+j)} \frac{\partial g_i(\boldsymbol{u})}{\partial u(k+p)}, \end{aligned} \tag{4.48}$$

where $j = 0, \ldots, N_u - 1$, $p = 0, \ldots, N_u - 1$. The partial derivatives in (4.48) can be calculated using (4.40).

4.2.7 *Model Linearization*

For nonlinear systems, a minimization of a cost function subject to constraints constitutes a nonlinear programming task, the solving of which can be troublesome. The main problem areas are as follows: (i) the multimodality of the cost, (ii) the computational burden of the nonlinear optimization procedure, (iii) problems with the procedure converging to an acceptable solution. These problems were discussed in our previous works [34, 39], where certain nonlinear programming algorithms were employed to realize MPC. In the following, an alternative solution is proposed.

In order to avoid the problems associated with nonlinear optimization, the approach based on instantaneous linearization of the neural network model is applied

[30, 35]. At each sampling time, the neural network (4.10) is linearized around the current operating point using expansion into the Taylor series. As a result, a linear model of the form:

$$\hat{y}_l(k+1) = \boldsymbol{\theta}\tilde{\boldsymbol{\varphi}}(k) + \hat{\varepsilon}(k) \tag{4.49}$$

is achieved, where $\boldsymbol{\theta} = [a_1, \ldots, a_{n_a}, b_1, \ldots, b_{n_b}]$ is the vector of feedback and feed-forward parameters, $\hat{\varepsilon}(k)$ is the integrated white noise. The extracted linear model aids in finding a solution to the optimization problem. Using a Diophantine equation, the prediction of the system output can be carried out in the following way:

$$\boldsymbol{y}_l = \boldsymbol{G}_l \Delta \boldsymbol{u} + \boldsymbol{f}_l, \tag{4.50}$$

where $\boldsymbol{y}_l = [\hat{y}_l(k+1), \ldots, \hat{y}_l(k+N_p)]^{\mathsf{T}}$, $\boldsymbol{G}_l$ is the matrix of step responses of the system (4.49) and $\boldsymbol{f}_l$ is the free response of the system (4.49). Taking into account that this procedure is well-documented, we omit the details of deriving $\boldsymbol{G}_l$ and $\boldsymbol{f}_l$. The interested reader is referred to [8, 47].

Applying the prediction formula (4.50), the optimization problem can be defined in the form of quadratic programming. In the case when no constraints are considered, the control law can be directly calculated as:

$$\Delta \boldsymbol{u} = -\boldsymbol{H}_l^{-1} \boldsymbol{g}_l, \tag{4.51}$$

where $\boldsymbol{H}_l = \boldsymbol{G}_l^{\mathsf{T}} \boldsymbol{G}_l + \rho \mathbf{I}$ ($\mathbf{I}$ stands for identity matrix), $\boldsymbol{g}_l = (\boldsymbol{y}_r - \boldsymbol{f}_l)^{\mathsf{T}} \boldsymbol{G}_l$. In such a case, the control applied to the system has the form:

$$u(k) = \Delta u(k) + u(k-1). \tag{4.52}$$

In the case when there are constraints imposed on the system, one needs to solve the respective constrained optimization problem. The constraints are formulated in the following general form:

$$\boldsymbol{A} \Delta \boldsymbol{u} \leq \boldsymbol{b}. \tag{4.53}$$

In the case when the input (4.5) and the output (4.7) constraints are considered, the augmented constraint representation takes the form:

$$\boldsymbol{A} = \begin{bmatrix} \boldsymbol{S} \\ -\boldsymbol{S} \\ \boldsymbol{G} \\ -\boldsymbol{G} \end{bmatrix}, \quad \boldsymbol{b} = \begin{bmatrix} \mathbf{1}(u_{max} - u(k-1)) \\ \mathbf{1}(-u_{min} + u(k-1)) \\ y_{max} - \boldsymbol{f} + \epsilon_{max} \\ -y_{min} + \boldsymbol{f} - \epsilon_{min} \end{bmatrix}, \tag{4.54}$$

where $\boldsymbol{S}$ is a lower triangular matrix with unit elements, $\mathbf{1} \in \mathbb{R}^{N_u}$ is the vector of ones, $\epsilon_{min} = [\epsilon_{min}(k+1), \ldots, \epsilon_{min}(k+N_y)]$, $\epsilon_{max} = [\epsilon_{max}(k+1), \ldots, \epsilon_{max}(k+N_y)]$. After substituting $\boldsymbol{y}_l$ for $\boldsymbol{y}$ in (4.1) the constrained optimization problem becomes:

$$\begin{aligned} (\Delta \boldsymbol{u}^{\mathsf{T}}, \boldsymbol{\epsilon}^{\mathsf{T}}) &\triangleq \arg\min J + \boldsymbol{\epsilon}^{\mathsf{T}}\boldsymbol{\epsilon} \\ \text{s.t.} \quad \Delta \boldsymbol{u}_c &= \boldsymbol{0} \\ \boldsymbol{A}\Delta \boldsymbol{u} &\leq \boldsymbol{b} \\ \boldsymbol{\epsilon} &\geq \boldsymbol{0} \end{aligned}, \tag{4.55}$$

where $\boldsymbol{\epsilon} = [\epsilon_{min}\ \epsilon_{max}]^{\mathsf{T}}$, $\Delta \boldsymbol{u}_c = [\Delta u(k+N_u), \ldots, \Delta u(k+N_p-1)]^{\mathsf{T}}$. The problem (4.55) can be solved using, e.g. the `quadprog` function of the Optimization toolbox in MATLAB®. Alternatively, the Hildreth's procedure as described in [49] is adopted. The Hildreth's algorithm is a realization of a dual method that can be used to identify constraints that are not active. These constraints are then eliminated in the solution. A dual method will lead to a very simple programming procedure, which is less complex than the original primal method.

4.3 Fault-Tolerant MPC

It should be pointed out that the interest in fault-tolerant control systems have grown in the last decade [6, 11, 31, 45, 46]: indeed, one observes focused efforts in designing control systems able to maintain current performance of the system as close as possible to the desirable one and to preserve stability conditions in the presence of unexpected changes caused by faults. In this context, MPC seems to be suitable for FTC, as the representation of both faults and control objectives is relatively simple. Some faults can be represented through the modification of constraints in the MPC algorithm, while others can be handled through modification of a system model [8, 17]. Except certain robust conservative solutions, MPC regardless its sophisticated operation idea and advantages, is not able to handle sensor or actuator faults. To develop fault-tolerant MPC (FT-MPC), the fundamental algorithm should be enriched not only with fault detection and isolation (FDI) unit but also with the supervisory one to properly manage the control algorithm based on information acquired from an FDI subsystem. This section is devoted to a description of a methodology of compensating sensor faults. When a fault is detected and isolated, the next step of the fault diagnosis is an estimation of the sensor fault size. Based on this information, the faulty measurements are corrected/reconstructed and subsequently passed on to the controller.

4.3.1 A Fault-Diagnosis Unit

A binary diagnostic matrix (BDM) [21, 22] is one of the most popular methods of fault diagnosis. It is a matrix of relations defined as the Cartesian product of the set of faults and diagnostic signals. Each relation can be defined by attributing to each diagnostic signal a subset of the faults that are detected by this signal.

Table 4.1 A binary diagnostic matrix

$r\backslash f$	f_1	f_2	f_3	f_4
r_1	0	0	1	1
r_2	1	0	1	1
r_3	0	1	0	0

The existence of a fault causes the appearance of a diagnostic signal, whose value equals 1. BDM can be determined on the using the residuals calculated taking into account a number of the so-called partial models of the diagnosed process (designed for the smallest possible parts of the process using the available measurements) with the set of partial models covering the entire system. Each residual (also referred to as a diagnostic signal) is sensitive to a specific group of faults. Thus, for each residual one can assign a specific set of faults. In order to obtain a fault signature, each residual should be evaluated. The simplest decision algorithm is to compare the absolute value of the residual with the threshold T:

$$s(r_i) = \begin{cases} 0 & \text{if } |r_i| \leq T, \\ 1 & \text{if } |r_i| > T. \end{cases} \tag{4.56}$$

To illustrate the underlying idea let us consider the set of three diagnostic signals:

$$\mathcal{F}(r_1) = \{f_3, f_4\}, \tag{4.57}$$

$$\mathcal{F}(r_2) = \{f_1, f_3, f_4\}, \tag{4.58}$$

$$\mathcal{F}(r_3) = \{f_2\}. \tag{4.59}$$

and the resulting binary diagnostic matrix (Table 4.1). It is evident that the residual r_1 is sensitive to faults f_3 and f_4, the residual r_2 is sensitive to f_1 and f_3 and r_3 sensitive to f_1 only. We can also observe that faults f_3 and f_4 are indistinguishable as they have the same fault signature. A remedy for improving fault distinguishability might lie in the application of a multi-valued evaluation instead of the binary one. Let us consider a three-value evaluation defined as follows:

$$s(r_i) = \begin{cases} 0 & \text{if } r_i \in [T_l, T_u], \\ +1 & \text{if } r_i > T_u, \\ -1 & \text{if } r_i < T_l, \end{cases} \tag{4.60}$$

where T_u and T_l represent the upper and lower thresholds, respectively, which are derived, for example, using a simple 3-standard deviation method. Assuming that the residual is $\mathcal{N}(m, \sigma)$ random variable, thresholds are assigned to the values:

$$T_u = m + 3\sigma, \quad T_l = m - 3\sigma. \tag{4.61}$$

The probability that the residual exceeds the threshold (4.61) is equal to 0.00135. Both, the mean value m and the standard deviation σ for each residual were calculated

Table 4.2 A multi-valued diagnostic matrix

$r \backslash f$	f_1	f_2	f_3	f_4
r_1	0	0	-1	$+1$
r_2	1	0	$+1$	$+1$
r_3	0	1	0	0

over the testing set recorded in closed-loop control in the steady state. Therefore, the original BDM represented by Table 4.1 might be potentially rewritten as a multi-valued diagnostic matrix (MDM) (see Table 4.2). Now, owing to the multi-valued evaluation, faults f_3 and f_4 become distinguishable. However, sometimes even if a multi-valued representation is used, some faults still cannot be isolated. Therefore, in order to improve the fault isolation quality, the dynamics of residuals can be taken into account. The possible solution is to investigate the order of the occurring symptoms or analyze the intensity of a fault [21]:

$$S(f_k) = \frac{1}{N} \sum_{j:r_j \in R(f_k)} \frac{r_j}{T_j}, \tag{4.62}$$

where $S(f_k)$ is the size of the fault f_k, $R(f_k)$ represents the set of residuals sensitive to the fault f_k, N is the window length determining the residual values taken to calculate the fault intensity, T_j stands for the threshold imposed on the j-th residual. Based on the above deliberations, we can say that the process of fault isolation consists of two steps. The first step is to explore the diagnostic matrix. The second step is to calculate the fault intensity to ensure that the faults are isolated. It should be emphasized that the second step is carried out only if faults are still indistinguishable.

4.3.2 *Sensor Fault Size Estimation*

Any automatic control system can hide faults from being observed, especially the faults of a small size. In our work [36], it was shown that predictive controllers based on neural networks possesses remarkable fault-tolerant control properties. Therefore, it is necessary to design the fault-diagnosis block providing information on abnormal working conditions of the plant. As far as actuator or process faults are concerned, predictive control can work in a required fashion, however sensor faults should be treated in a different way. For example, if we consider a fault occurring in a sensor measuring the water level in a tank, the control system immediately reacts to the change in the system output, which is the direct consequence of the wrongly measured water level in the boiler. In that case, the control signal value should be kept the same as before the fault occurrence, because such a fault does not change the water level in the boiler. Thus, an estimation of the size of a sensor fault is very important when designing fault-tolerant predictive control.

The multi-valued diagnostic matrix described in the previous section allows for an isolation of faults. When a fault is isolated, the next stage is fault size estimation.

This is a necessary condition for ensuring the proper work of the control system. Let us assume that the fault f_k has been isolated:

$$f_k \neq 0, \quad f_i = 0, \quad i = 1, \ldots, K, \quad i \neq k, \tag{4.63}$$

where K is the number of faults considered. All residuals sensitive to the fault f_k form the set

$$R(f_k) = \{r_j \in R : r_j \neq 0\}, \tag{4.64}$$

where R is the set of all residuals. In order to calculate the size of a sensor fault, the set of residuals sensitive to the fault f_k should be further reduced to a subset consisting of residuals defined using process variable measured by the isolated sensor:

$$\bar{R}(f_k) = \{r_j \in R_k : r_j \neq 0\}, \tag{4.65}$$

where R_k is a set of residuals sensitive to the fault f_k and determined using measurements acquired by the isolated sensor. Thus, the size of the fault can be directly calculated numerically as the value of a residual sensitive to the fault f_k, when only one residual is sensitive to this fault, or as the mean value of all residuals belonging to $\bar{R}(f_k)$, in case the set $\bar{R}(f_k)$ contains more than one residual.

4.4 An Experimental Study — A Tank Unit

4.4.1 A Tank Unit

The technological process considered here is a laboratory installation developed at the Institute of Automatic Control and Robotics of the Warsaw University of Technology (Fig. 4.1). The installation is dedicated for the investigation of various diagnostic methods for industrial actuators and sensors [36, 37]. The system includes a tank, a storage vessel, a control valve with a positioner, a pump, and transducers to measure process variables. The tank is realized in the form of a horizontally placed cylinder, which introduces strong nonlinearity into the static characteristic of the system. The laboratory stand was implemented in MATLAB®/Simulink® software. The simulation model was successfully verified using real data acquired from the physical installation. Figure 4.2 shows the scheme of the tank unit where the process variables that can be measured are marked, while the description of the process variables is presented in Table 4.3. The advantage of the simulator over the real process is that the simulator makes it possible to check the behaviour of the proposed predictive control in the presence of a wide range of possible fault scenarios. The specification of faults is presented in Table 4.4. The considered scenarios are different, including both additive and multiplicative faults. Faults in different parts of the installation are proposed, including sensor, actuator and component faults. Faults placement is marked in Fig. 4.2. In short, the proposed set of faults renders it

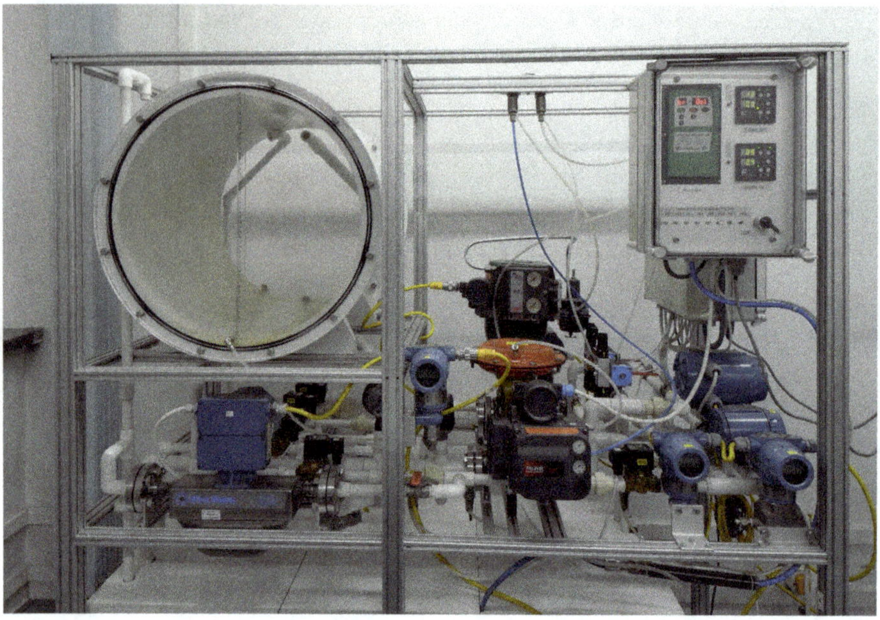

Fig. 4.1 A real-life laboratory installation [34]. ©2015 IEEE. Reproduced with permission

Fig. 4.2 A block scheme of the tank unit and possible fault placement [34]. ©2015 IEEE. Reproduced with permission

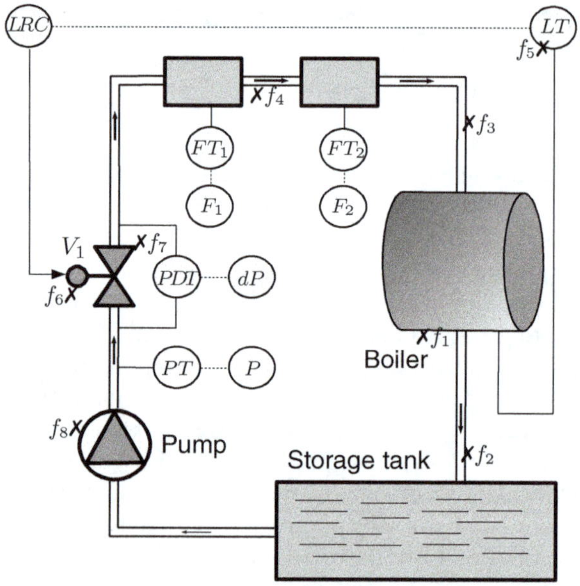

possible to investigate the widest range of fault-tolerance properties of the proposed predictive controller. Simulations are performed with the sample time T_s equal to 0.05 s.

Table 4.3 The specifications of process variables

Variable	Specification	Range
CV	Control value	0–100 %
dP	Pressure difference on valve V_1	0–275 kPa
P	Pressure before valve V_1	0–500 kPa
F_1	Flow (electromagnetic flowmeter)	0–5 m^3/h
F_2	Flow (Vortex flowmeter)	0–5 m^3/h
L	Water level in boiler	0–0.5 m

Table 4.4 The specification of faulty scenarios [34]. ©2015 IEEE. Reproduced with permission

Fault	Description	Type
f_1	Leakage from boiler	Additive (+0.083)
f_2	Outflow choking	Partly closed (50%)
f_3	Loss of internal pipe diameter	Partly closed (50%)
f_4	Leakage from pipe	Additive (−1.67)
f_5	Level transducer failure	Additive (−0.05)
f_6	Positioner fault	Multiplicative (0.7)
f_7	Valve head or servo-motor fault	Multiplicative (0.8)
f_8	Pump productivity reduction	Multiplicative (0.8)

4.4.2 Plant Modelling

In order to build a model of the process, a recurrent neural network of the NOE type was applied (as described in Chap. 2, Sect. 2.3.1). The model input was the control value (CV) and the model output was the level in the boiler (L). The first stage of modelling is data collecting. Training data were collected in the open-loop control. The tank unit is the process with slow dynamics. Filling the tank to the level equal to 0.25 m using the maximal input flow takes approximately 300 s. Therefore to design the tank model training data were sampled using the sampling time equal to 0.5 s. The input signal should be as much informative as possible. It means that it should be persistently exciting of a certain order, i.e. it should contain sufficiently many distinct frequencies. Based on the analysis carried out in [36], the input signal was chosen as random steps with levels from the interval (0, 100), each step lasting 200 s. Additionally, white noise of the magnitude of 10 was added to this signal. The input and output training data are presented in Fig. 4.3. The model structure was selected taking into account a compromise between the model complexity and quality [33]. As a results of a number of trails, the best neural model was found to have 7 hidden neurons with hyperbolic tangent activation functions, 1 linear output neuron, the input delay equal to 2 and the output delay equal to 2. The training sequence consisted of 5000 samples. The training process was carried out *off-line*

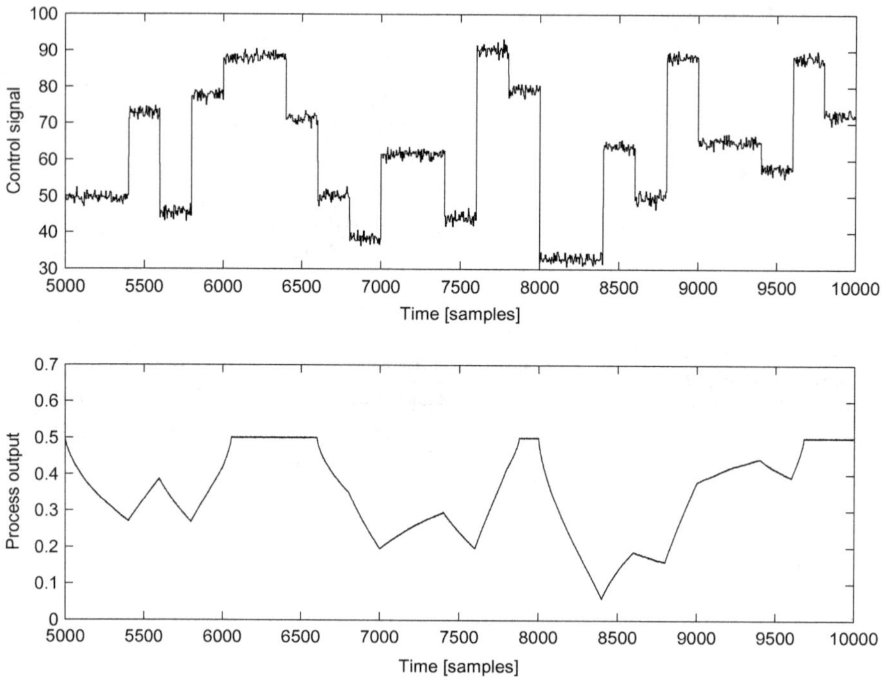

Fig. 4.3 The training data: the input (upper graph), and the output signals (lower graph)

for 200 steps with the Levenberg–Marquardt algorithm. In this setting, the k-step ahead prediction is achieved by recursively developing a 1-step ahead predictor. In order to measure the quality of the modelling, the sum of squared errors (SSE) and mean squared error (MSE) are employed:

$$\mathrm{SSE}(a, b) = \sum_{i=1}^{N} (a(i) - b(i))^2, \tag{4.66}$$

$$\mathrm{MSE}(a, b) = \frac{1}{N} \sum_{i=1}^{N} (a(i) - b(i))^2, \tag{4.67}$$

where N is the number of samples, and a and b stand for time series. Quality indexes for selected k-step ahead predictors are presented in Table 4.5. Although it is observed that the uncertainty of the 1-step ahead predictor affects recursively the k-step ahead predictors developed, the 15-step ahead predictor predicts the behaviour of the process to an acceptable level (Fig. 4.4).

Table 4.5 The quality indexes of the predictors [34]. ©2015 IEEE. Reproduced with permission

Quality index	k-step ahead predictor					
	$k = 1$	$k = 3$	$k = 5$	$k = 7$	$k = 10$	$k = 15$
SSE(y, $\hat{y}$)	10.84	11.44	12.68	14.64	18.61	26.46
MSE(y, $\hat{y}$)	$5.42 \cdot 10^{-4}$	$5.72 \cdot 10^{-4}$	$6.34 \cdot 10^{-4}$	$7.32 \cdot 10^{-4}$	$9.3 \cdot 10^{-4}$	$1.3 \cdot 10^{-3}$

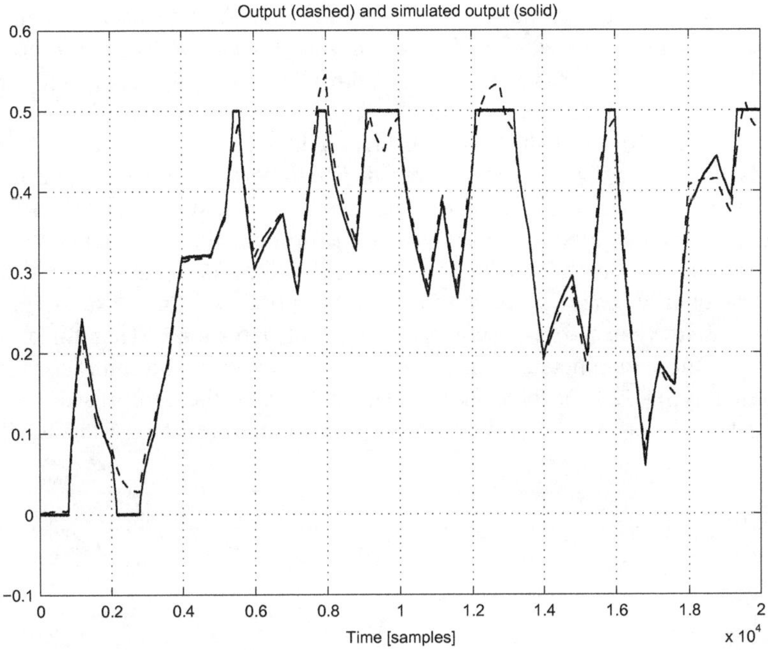

Fig. 4.4 The 15-step ahead predictor testing

4.4.3 Control

The controller parameters are set as follows: $N_p = 15$, $N_u = 2$, $\rho = 2 \cdot 10^{-5}$. Such experiment settings were found using a trial-and-error procedure and ensure a decent performance of the control system. The control sequence is constrained with the upper control bound $u_{max} = 100$ (the maximal value generated by the actuator), and the lower control bound $u_{min} = 20$ to cope with the problem of dead-zone of the boiler (when the control signal has the value less than 40 the outflow is greater than the inflow). In the view of Theorem 4.1, the constraint horizon $N_c = 3$. The penalty coefficients of the cost function (4.34) were set as follows: $\mu = 1$ and $\lambda = 1$. Additionally, a disturbance model (4.12) is used in order to consider the unmeasured disturbances affecting the system. The fact that a disturbance model is used means that the control system becomes robust in certain ways, as shown by the experiments described in the subsequent parts of the chapter. After a number of trials, the value of the parameter k_c was set to 0.4. With this setting, a decent level of the convergence of the tracking error to zero can be achieved. The performance of the proposed stable predictive control in the presence of a disturbance (white noise with zero mean and standard deviation $\sigma = 0.001$) is shown in Fig. 4.5. The plant output follows the reference almost immediately (Fig. 4.5a). An introduction of terminal equality constraints smooths the control signals (Fig. 4.5b). We note that the reference tracking is performed with a decent quality ($\mathrm{SSE}(y_r, y) = 73.56$) and even a more intensive disturbance (white noise with σ=0.01) does not significantly reduce the control quality ($\mathrm{SSE}(y_r, y) = 75.30$). The proposed control scheme is also compared with the PID controller. For the PID controller, the control quality was about 7% worse than for stable MPC. Moreover, the PID controller could not guarantee zero tracking error in steady-state when the setpoint is changed. The important problem here is to select proper values of μ and λ. Using larger values for these penalty parameters ensures an acceptable approximation to the solution of the optimization problem (4.45). However, large values of μ and λ mean that the penalty terms dominate the cost and that the control quality has not such an impact on the optimal solution. Excessively large values of μ and λ can give rise to overshots in the plant response. Therefore, a proper selection of these parameters is only a compromise between the control quality and the optimality of the control sequence. The stability of the control system is also investigated. Owing to the application of the terminal constraints, the stability of

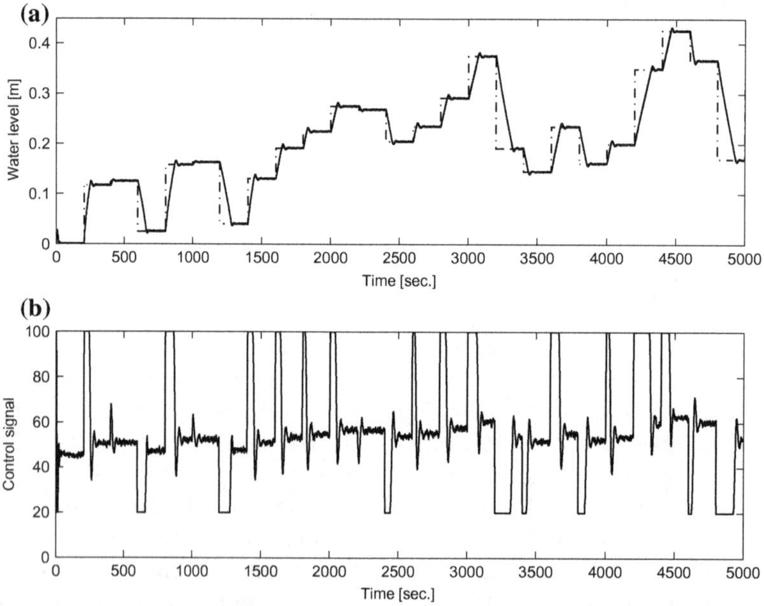

Fig. 4.5 Process output (solid) and reference (dashed) (**a**), the control signal (**b**) [34]. ©2015 IEEE. Reproduced with permission

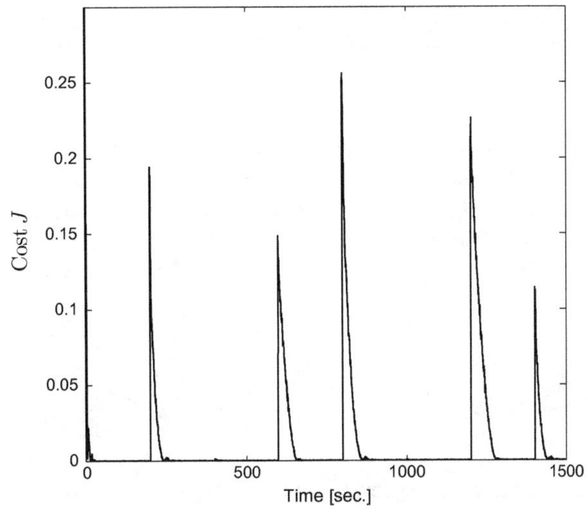

Fig. 4.6 The evolution of the cost function J [34]. ©2015 IEEE. Reproduced with permission

the loop and the monotonicity of the cost function J are observed, which confirms that the proposed predictive control systems guarantees stability (Fig. 4.6). The fast convergence of the cost to zero can be ensured by setting relatively large values of penalty factors λ and μ. However, as pointed out previously, excessively large values of μ and λ can cause an overshot in the plant response and, consequently, result perturbations in terms of the monotonicity of the cost.

4.4.4 Fault Diagnosis

The following five single-input single-output partial models are considered [34, 37]:

- the pump model:

$$\hat{P} = f_{m1}(F_1), \tag{4.68}$$

- the tank model:

$$\hat{L}_1 = f_{m2}(F_1), \tag{4.69}$$

- the valve model:

$$\hat{F}_1 = f_{m3}(dP), \tag{4.70}$$

- the positioner model:

$$\hat{F}_2 = f_{m4}(CV), \tag{4.71}$$

- the system model:

$$\hat{L}_2 = f_{m5}(CV), \tag{4.72}$$

where $f_{m1}(\cdot)$, $f_{m2}(\cdot)$, $f_{m3}(\cdot)$, $f_{m4}(\cdot)$ and $f_{m5}(\cdot)$ stand for nonlinear mappings between input and output variables for each single partial model, respectively, and $\hat{F}_1$ and $\hat{F}_2$ are estimated flows given by the valve and positioner models and $\hat{P}$ is the pressure estimated by the pump model. Based on these partial models, the following residuals are calculated:

$$\begin{aligned} r_1 &= P - f_{m1}(F_1), \\ r_2 &= L - f_{m2}(F_1), \\ r_3 &= F_1 - f_{m3}(dP), \\ r_4 &= F_1 - f_{m4}(CV), \\ r_5 &= L - f_{m5}(CV). \end{aligned} \tag{4.73}$$

Thus, based on expert knowledge about the considered process for each residual one can assign the specific set of faults as follows:

$$\mathcal{F}(r_1) = \{f_3, f_4, f_8\}, \tag{4.74}$$
$$\mathcal{F}(r_2) = \{f_1, f_2, f_3, f_4, f_5, f_6, f_7, f_8\}, \tag{4.75}$$
$$\mathcal{F}(r_3) = \{f_3, f_4, f_8\}, \tag{4.76}$$
$$\mathcal{F}(r_4) = \{f_3, f_4, f_5, f_6, f_7, f_8\}, \tag{4.77}$$
$$\mathcal{F}(r_5) = \{f_1, f_2, f_3, f_4, f_5, f_6, f_7, f_8\}. \tag{4.78}$$

The binary diagnostic matrix that corresponds to the sets (4.74)–(4.78) is shown in Table 4.6. Each fault is represented by the fault signature, which is a vector of diagnostic signal values that correspond to this fault. The first observation is that using the proposed methodology it is not possible to distinguish between the faults f_1 and f_2. Moreover, f_3 has the same fault signature as f_4 and f_8. Finally, the faults f_5, f_6 and f_7 are also indistinguishable. Therefore, a multi-valued evaluation was applied with MDM as described in Table 4.7. Analyzing Table 4.7, it can be observed that the residuals r_2 and r_5 provide the same information for MDM. However, they can still provide important data on faults, especially taking into account the dynamics of the residuals. Moreover, there still exist faults that cannot be isolated (pairs $\{f_3, f_8\}$, and $\{f_6, f_7\}$). The solution was to consider the intensity of a fault (see the last row of Table 4.9 for details).

Table 4.6 A binary diagnostic matrix

$r \setminus f$	f_1	f_2	f_3	f_4	f_5	f_6	f_7	f_8
r_1	0	0	1	1	0	0	0	1
r_2	1	1	1	1	1	1	1	1
r_3	0	0	1	1	0	0	0	1
r_4	0	0	1	1	1	1	1	1
r_5	1	1	1	1	1	1	1	1

Table 4.7 A multi-valued diagnostic matrix [34]. ©2015 IEEE. Reproduced with permission

$r \setminus f$	f_1	f_2	f_3	f_4	f_5	f_6	f_7	f_8
r_1	0	0	−1	−1	0	0	0	−1
r_2	−1	+1	+1	−1	−1	+1	+1	+1
r_3	0	0	−1	−1	0	0	0	−1
r_4	0	0	−1	−1	−1	−1	−1	−1
r_5	−1	+1	+1	−1	−1	+1	+1	+1

Table 4.8 The model specification [34]. ©2015 IEEE. Reproduced with permission

Model	v	n_a	n_b	σ_h	σ_o	T_s[s]	T_l	T_u
Pump model	3	1	1	tanh	Linear	0.05	−0.0072	0.0178
Boiler model	10	2	2	tanh	Linear	0.05	−0.0023	−0.000195
Valve model	5	2	1	tanh	Linear	0.05	−0.0166	0.007
Positioner model	5	1	1	tanh	Linear	0.05	−0.0245	0.0093

Partial models (4.68)–(4.71) were designed by means of the recurrent neural network (4.10). First, the modelling was carried out in the open-loop control, however, the quality of the models achieved in this way, as tested in the closed-loop control, was not satisfactory. Therefore, the modelling was performed in the closed-loop control. Training data were recorded during the normal operation of the system using a stable predictive controller with settings presented in Sect. 4.4.3. The reference signal used was in the form of random steps sequence with levels from the interval (0, 0.5); each step lasting 200 s. Since the obtained values were relatively large, the pressure before valve P as well as the pressure difference dP signals were scaled linearly to fall into the range [0, 1]. The structures of neural models were selected experimentally using the same criteria as presented in Sect. 4.4.2. All neural models had only one hidden layer. The specification of the best models is presented in Table 4.8. The training was carried out off-line using a recurrent version of the Levenberg–Marquardt algorithm [29]. The fault diagnosis is performed by analyzing the residuals calculated using partial models. The threshold values, derived according to (4.61), assigned to each partial model and required in terms of decision making are presented in the last two columns of Table 4.8. The number of false alarms caused by temporary fault diagnosis should be as small as possible. In order to determine the quality of the fault detection, the detection time t_d is used, defined as a period of time measured from the fault start-up time to a permanent, true decision about a fault [21, 34]. To be sure that a residual permanently exceeds thresholds, a time window with the length 1 s (20 samples) was used. If the residual exceeds a threshold for a period of time >1 s, then a fault is detected and signalled. Such a procedure is applied to each residual. The results of the fault diagnosis for the considered set of faults are presented in Table 4.9. Each fault was introduced in the experiment at 500 s. Analyzing the results, it is obvious that by using the proposed approach all faults were definitely detected. The second stage is the fault isolation. If a fault is detected, the

Table 4.9 The results of the fault diagnosis [34]. ©2015 IEEE. Reproduced with permission

Index	Fault scenario							
	f_1	f_2	f_3	f_4	f_5	f_6	f_7	f_8
$t_d[s]$	11.45	2.3	1	1	1	1	1	1
$t_i[s]$	11.45	9.05	20.4	1	2.85	1	1	3.05
Isolated as	f_1	f_2	f_3	f_4	f_5	f_6	f_7	f_8
$S(f_k)$	–	–	337.65	–	–	42	32.83	110.11

diagnostic system tries to locate it. This is done simply by searching the diagnostic table (Table 4.7), in order to find a match for the fault signature. The fault signature is said to be permanent, if it does not change within a time window. Similar to the case of the fault detection, in order to isolate faults a time window with the length 1 s was applied. The fault isolation quality is determined by means of the so-called isolation time t_i, which is defined as a period of time measured from the beginning of fault start-up time to the moment of fault isolation [21, 34]. The results of the fault isolation are presented in the third row of Table 4.9. Isolating certain faults can be troublesome. Even if a fault was detected relatively fast, its isolation can take a long time, as is the case of e.g. the fault f_3. Some faults are not distinguishable. In the considered case, the faults f_3 and f_8 as well as f_6 and f_7 possess the same fault signature, as can be seen by comparing the third column with the eight one, and the seventh and the eighth column in Table 4.9. This problem can be solved using the notion of a fault intensity, as discussed in Sect. 4.3.1. The fault intensity is calculated using the same time window as in the case of both fault detection and fault isolation. Example fault intensities are presented in the last row of Table 4.9. It is clear that using the notion of a fault intensity one can definitely distinguish f_3 from f_8, and f_6 from f_7, and vice versa.

4.4.5 Fault Tolerance

Table 4.10 includes the fault tolerance results for MPC without a disturbance model (the first row—MPC), MPC with the disturbance model (the second row—MPCD) and the classical PID (the third row—PID). Each considered fault was simulated

Table 4.10 The fault tolerance results [34]. ©2015 IEEE. Reproduced with permission

Method	Fault scenario				
	f_1	f_2	f_4	f_6	f_7
MPC	0.00024	0.00865	0.00389	0.0499	0.0166
MPCD	0.00024	0.00459	0.00039	0.0013	0.00057
PID	0.00027	0.3002	0.05089	0.2218	0.0748
	f_8	$f_2 + f_4$	$f_2 + f_7$	$f_4 + f_7$	$f_7 + f_8$
MPC	0.00127	–	–	–	–
MPCD	0.00026	0.00484	0.00476	0.00187	0.0016
PID	0.00605	–	–	–	–

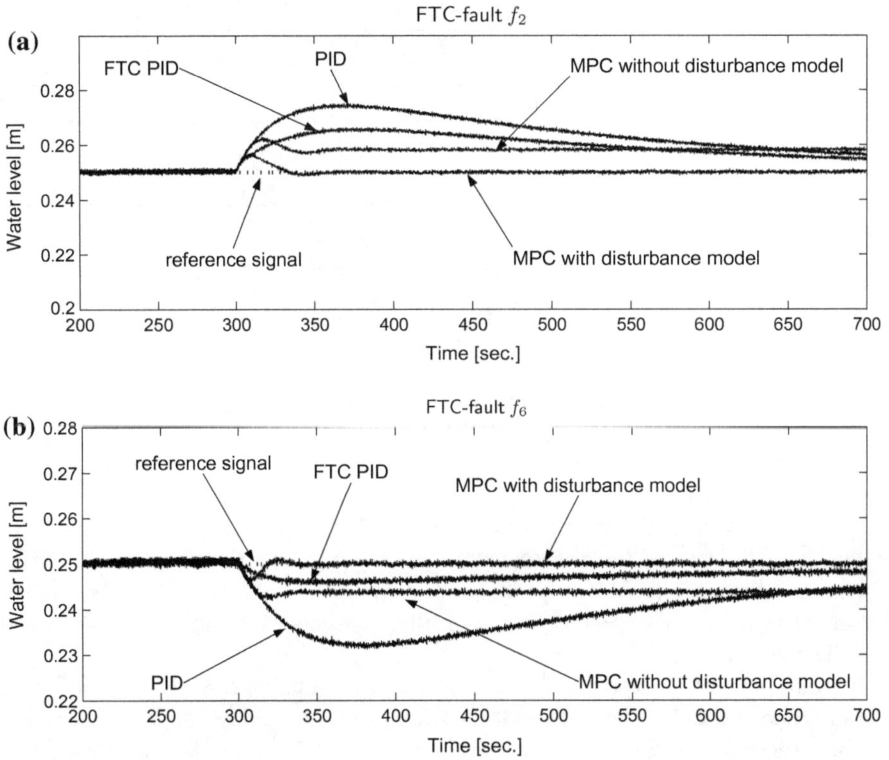

Fig. 4.7 Fault tolerance: the fault f_2 (**a**), and f_6 (**b**) [34]. ©2015 IEEE. Reproduced with permission

at 300 s with the experiment lasting 700 s. For comparison, for the nominal work of the control system $\text{SSE}(y_r, y) = 0.0002$. Two examples of fault-tolerant control are shown in Fig. 4.7a (fault f_2), and b (fault f_6). The only fault that was not compensated is f_3. In that case, it was observed that the control system tried to keep the required level of water but after $\tilde{8}0$ s from the fault start-up time, due to limited maximum inflow to the tank, which is less than the outflow, the fault effect cannot be compensated in any way. For MPC without the disturbance model, the fault effect changed the dynamic properties of the plant, whereas the predictive controller relies on the model developed for normal operating conditions. In such cases, a prediction of the output can be inaccurate and some error in the steady-state can be observed (Fig. 4.7). A much better behaviour is observed in the case of MPC that uses the disturbance model. As can be seen in Table 4.10, almost all the process/actuator faults (excluding f_3) are compensated with steady-state error near zero. The PID controller is able to somehow compensate faults but the compensation time is quite long (see Fig. 4.7). The PID controller does not respond to the changing conditions of the system as fast as the predictive controller, therefore the quality indexes for the PID controller presented in Table 4.10 are not as good as for the MPCD controller.

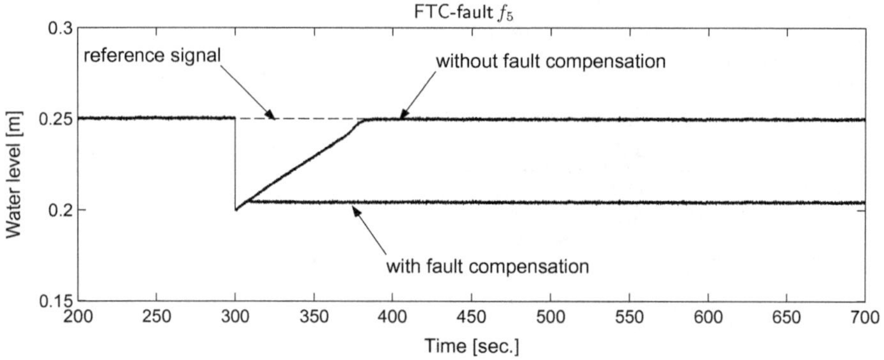

Fig. 4.8 Fault tolerance: the fault f_5 [34]. ©2015 IEEE. Reproduced with permission

The proposed control scheme is also compared with a different fault-tolerant approach of the passive type [6, 10], where reconfiguration of the controller is simply done by switching between the already designed PID controllers. Each controller is designed for one of the predefined faulty situation. The work of such a FTC system is illustrated in Fig. 4.7 (curves marked as FTC PID). FTC PID exhibits better performance than the fixed PID controller but is not as effective as the proposed MPCD controller.

A multiple fault case is also investigated. Two faults were simulated in a row, the first one at the 300 s mark and the second one at the 350 s mark. This portion of the experiment was carried out only for the MPCD controller and the results are presented in the last four columns of Table 4.10. Even in the case of multiple faults, the control system is able to operate properly and faults are relatively quickly compensated.

In the case of the sensor fault f_5, a control signal should be kept the same as before the fault occurrence, because such a fault does not change the water level in the tank. A reconstruction of the water level is done using the estimated level $\hat{L}$ at the moment of sensor fault isolation. The water level in the tank is estimated using the tank model ($\hat{L}_1 = f_{m2}(F_1)$). The difference (dL) between the measured boiler level and the output of the tank model at the moment of fault isolation was equal to $dL = -0.0459$. The system behaviour in the case of sensor fault compensation is depicted in Fig. 4.8. It is observed that the wrongly measured value of the tank level is changed to a proper value and transferred to the predictive controller, not affecting the control signal in a significant way. Consequently, the predictive controller does not try to increase the water level to the reference value.

4.5 Robust MPC

The robustness of a control system is described in reference to a specific uncertainty range and specific stability and performance criteria. In spite of rich literature devoted to robust control in the context of linear systems, very little is known about robust control of linear systems with constraints or the case of nonlinear systems. In general, the model uncertainty in the case of robust control can have two main sources [4]: (i) an unmodelled dynamics of a plant, (ii) an unmeasured noise/disturbances entering the plant. However, in the framework of linear time-invariant systems, several different approaches have been proposed, e.g. impulse/step responses, a polytopic uncertainty or bounded input disturbances. Generally speaking, the existing methods can be divided into two classes: structured and unstructured uncertainties [44]. These uncertainty descriptions, however, are very useful in the case of linear time-invariant systems, especially using the H_∞ paradigm. An intuitive method for achieving the robustness is to solve the min-max problem [19]. However, such an approach leads to very time-consuming algorithms; the problems are especially noticeable in the case of nonlinear systems. A remedy for this could be related to an application of approximate reachable sets determined by means of the interval arithmetic [24]. A drawback of this solution is that the constraints are formulated using uncertain evolution sets, which can complicate the optimization process. Recently, neural networks have been applied to robust MPC (RMPC) synthesis [23, 50]. In these works, however, recurrent neural networks are used to solve the min-max optimal control problem. A possible solution is to use neural networks to estimate uncertainty of the model used in the framework of predictive control. In such a situation, the approach is to use dynamic neural networks to achieve two goals: to derive the nominal model of a plant and to deal with the uncertainty associated with such a model. This approach is motivated by the fact that in many cases a controlled plant is poorly damped or has small stability margin and data representing the plant have to be recorded in the closed-loop control with a preliminary selected controller. Consequently, the quality of the model will be limited to the properties of the controller used to record the data. It follows that the model should be improved in some way. To that end the MEM method, introduced in Chap. 2, Sect. 2.5.2, is applied [35, 39].

To simplify matters, let us assume that an error model has been already designed. Similarly as in the case of the fundamental model of the plant (4.10), at each sampling time the error model (4.20) is linearized giving:

$$\hat{w}_l(k+1) = \boldsymbol{\theta}_w \phi(k) + \hat{\varepsilon}(k), \tag{4.79}$$

where $\boldsymbol{\theta}_w = [c_1, \ldots, c_{n_c}, d_1, \ldots, d_{n_d}]$ is the vector of feedback and feed-forward parameters, $\hat{\varepsilon}(k)$ is the integrated white noise. Based on the linearized error model (4.79), the prediction equation is derived as follows:

$$\boldsymbol{w}_l = \boldsymbol{G}_w \Delta \boldsymbol{u} + \boldsymbol{f}_w, \tag{4.80}$$

where the matrix $\boldsymbol{G}_w$ and the vector $\boldsymbol{f}_w$ are the matrix of step responses and the free response of the error model (4.79), respectively. Taking into account (4.21) and (4.50), predictions of the linearized robust model can be expressed as:

$$\bar{\boldsymbol{y}}_l = (\boldsymbol{G}_l + \boldsymbol{G}_w)\Delta \boldsymbol{u} + (\boldsymbol{f}_l + \boldsymbol{f}_w) = \tilde{\boldsymbol{G}}\Delta \boldsymbol{u} + \tilde{\boldsymbol{f}}, \tag{4.81}$$

where $\bar{\boldsymbol{y}}_l = [\bar{y}_l(k+1), \dots, \bar{y}_l(k+N_p)]^{\mathsf{T}}$. Based on the prediction equation (4.81), a robust form of predictive control is proposed. Robustness of MPC can be achieved by (i) modifying the cost function, (ii) a proper definition of the output constraints. The realisation of the former is related to the use of the vector of future predictions of the robust model $\bar{\boldsymbol{y}}_l$ instead of the vector of future predictions of the nominal model $\hat{\boldsymbol{y}}$. The cost function becomes:

$$\bar{J} = (\boldsymbol{y}_r - \bar{\boldsymbol{y}}_l)^{\mathsf{T}}(\boldsymbol{y}_r - \bar{\boldsymbol{y}}_l) + \rho\Delta\boldsymbol{u}^{\mathsf{T}}\Delta\boldsymbol{u}. \tag{4.82}$$

As the model (4.81) is employed, the optimization problem reduces to quadratic programming problem, which for the unconstrained form can be solved using (4.51), where $\boldsymbol{H} = \tilde{\boldsymbol{G}}^{\mathsf{T}}\tilde{\boldsymbol{G}} + \rho\mathbf{I}$, $\boldsymbol{g} = (\boldsymbol{y}_r - \tilde{\boldsymbol{f}})^{\mathsf{T}}\tilde{\boldsymbol{G}}$, and then control can be determined using (4.52). For the constrained form, the problem can be formulated analogously to (4.55). Robust output constraints can be formulated taking into account the bounds imposed on the system outputs:

$$y_{min} \leq \bar{y}_l(k+i) \leq y_{max}, \quad i = 1, \dots, N_y. \tag{4.83}$$

Owing to the same reasons as in the case of the output constraints, we can use soft constraints as follows:

$$y_{min} - \bar{\epsilon}_{min}(k+i) \leq \bar{y}_l(k+i) \leq y_{max} + \bar{\epsilon}_{max}(k+i), \quad i = 1, \dots, N_y. \tag{4.84}$$

Moreover, taking into account the uncertainty bounds (4.22) and (4.23), the output constraints imposed on the fundamental model of the system can be redefined. As a result we obtain:

$$\begin{aligned} y_{min} - \underline{w}(k+i) - \epsilon_{min}(k+i) &\leq \tilde{y}_l(k+i) \\ &\leq y_{max} - \overline{w}(k+i) + \epsilon_{max}(k+i), \ i = \dots, N_y. \end{aligned} \tag{4.85}$$

Using the prediction formula (4.81), the constraint (4.84) can be rewritten in the following way:

$$y_{min} - \tilde{\boldsymbol{f}} - \bar{\boldsymbol{\epsilon}}_{min} \leq \tilde{\boldsymbol{G}}\Delta\boldsymbol{u} \leq y_{max} - \tilde{\boldsymbol{f}} + \bar{\boldsymbol{\epsilon}}_{max}, \tag{4.86}$$

where $\underline{\boldsymbol{w}} = [\underline{w}(k+1), \dots, \underline{w}(k+N_y)]^{\mathsf{T}}$ and $\overline{\boldsymbol{w}} = [\overline{w}(k+1), \dots, \overline{w}(k+N_y)]^{\mathsf{T}}$, $\bar{\boldsymbol{\epsilon}}_{min} = [\bar{\epsilon}_{min}(k+1), \dots, \bar{\epsilon}_{min}(k+N_y)]$, $\bar{\boldsymbol{\epsilon}}_{max} = [\bar{\epsilon}_{max}(k+1), \dots, \bar{\epsilon}_{max}(k+N_y)]$. Then

$$\bar{\boldsymbol{A}} = \begin{bmatrix} \boldsymbol{S} \\ -\boldsymbol{S} \\ \boldsymbol{G}_l \\ -\boldsymbol{G}_l \\ \tilde{\boldsymbol{G}} \\ -\tilde{\boldsymbol{G}} \end{bmatrix}, \quad \bar{\boldsymbol{b}} = \begin{bmatrix} \mathbf{1}(u_{max} - u(k-1)) \\ \mathbf{1}(-u_{min} + u(k-1)) \\ y_{max} - \overline{\boldsymbol{w}} - \boldsymbol{f}_l + \boldsymbol{\epsilon}_{max} \\ -y_{min} + \underline{\boldsymbol{w}} + \boldsymbol{f}_l + \boldsymbol{\epsilon}_{min} \\ y_{max} - \tilde{\boldsymbol{f}} + \bar{\boldsymbol{\epsilon}}_{max} \\ -y_{min} + \tilde{\boldsymbol{f}} + \bar{\boldsymbol{\epsilon}}_{min} \end{bmatrix}, \tag{4.87}$$

and the proposed robust MPC takes the form:

$$\begin{aligned}(\Delta\boldsymbol{u}^{\mathsf{T}}, \boldsymbol{\epsilon}^{\mathsf{T}}, \bar{\boldsymbol{\epsilon}}^{\mathsf{T}}) &\triangleq \arg\min \bar{J} + \boldsymbol{\epsilon}^T\boldsymbol{\epsilon} + \bar{\boldsymbol{\epsilon}}^T\bar{\boldsymbol{\epsilon}} \\ \text{s.t.}\quad \Delta\boldsymbol{u}_c &= \mathbf{0} \\ \bar{\boldsymbol{A}}\Delta\boldsymbol{u} &\leq \bar{\boldsymbol{b}} \\ \boldsymbol{\epsilon} \geq \mathbf{0}&, \bar{\boldsymbol{\epsilon}} \geq \mathbf{0}\end{aligned}, \tag{4.88}$$

where $\bar{\boldsymbol{\epsilon}} = [\bar{\boldsymbol{\epsilon}}_{min}\ \bar{\boldsymbol{\epsilon}}_{max}]^{\mathsf{T}}$. The design procedure of robust MPC based on two-stages modelling by means of neural networks is summarized by the Algorithm 4.2.

Algorithm 4.2 Robust MPC based on two-stages modelling

Step 1. Record data $\{u(i), y(i)\}_{i=1}^{N}$ in closed-loop control with a classical PID controller.

Step 2. Design the fundamental model of a plant (4.10).

Step 3. Run MPC according to scheme (4.55) with multiplicative input and output uncertainty and once again record data $\{u(i), y(i)\}_{i=1}^{N}$.

Step 4. Using recorded data and the model (4.10), derive data $\{u(i), w(i)\}_{i=1}^{N}$ and identify an error model (4.19).

Step 5. Construct a robust model (4.81) and run robust MPC according to scheme (4.88) and evaluate its quality.

4.5.1 Robust Stability

One of the desired characteristics of robust control is what is known as robust stability. In this section, we investigate certain aspects of the asymptotic stability of the proposed robust control scheme by checking the monotonicity of the cost, as originally proposed in [42]. Among a number of methods used to achieve a stable MPC, the notion of terminal constraints is very popular:

$$e(k + N_p + j) = 0, \quad \forall j \in [1, N_c], \tag{4.89}$$

where $e(k+i) = y_r(k+i) - \bar{y}_l(k+i)$ is the tracking error, N_c is the constraints horizon. Clearly, it is assumed that beyond N_c the output of the model is the same as the reference at the end of the prediction horizon. It means that the tracking error beyond the prediction horizon is zero and the system is forced to the steady state. Let us revisit the robust MPC (4.88) but with the hard output constraints:

$$\tilde{\boldsymbol{A}} = \begin{bmatrix} \boldsymbol{S} \\ -\boldsymbol{S} \\ \boldsymbol{G}_l \\ -\boldsymbol{G}_l \\ \tilde{\boldsymbol{G}} \\ -\tilde{\boldsymbol{G}} \end{bmatrix}, \quad \tilde{\boldsymbol{b}} = \begin{bmatrix} \boldsymbol{1}(u_{max} - u(k-1)) \\ \boldsymbol{1}(-u_{min} + u(k-1)) \\ y_{max} - \overline{\boldsymbol{w}} - \boldsymbol{f}_l \\ -y_{min} + \underline{\boldsymbol{w}} + \boldsymbol{f}_l \\ y_{max} - \tilde{\boldsymbol{f}} \\ -y_{min} + \tilde{\boldsymbol{f}} \end{bmatrix} \tag{4.90}$$

and define $\boldsymbol{e} = [e(k+N_p+1), \ldots, e(k+N_p+N_c)]^\mathsf{T}$. Thus, stable robust MPC is represented as:

$$\begin{aligned} \Delta\boldsymbol{u} &\triangleq \arg\min \bar{J} \\ \text{s.t.} \quad \Delta\boldsymbol{u}_c &= \boldsymbol{0} \\ \tilde{\boldsymbol{A}}\Delta\boldsymbol{u} &\leq \tilde{\boldsymbol{b}} \\ \boldsymbol{e} &= \boldsymbol{0} \end{aligned} \quad . \tag{4.91}$$

Theorem 4.3 ([35]) *The robust model predictive control system (4.91) using a predictor based on (4.81) is asymptotically stable, in undisturbed case, if the following conditions are satisfied:*

1. $\rho \neq 0$,
2. $n_c = 0$,
3. $N_c = \max\left[n_a + 1, n_b + 1 + N_u - N_p, n_d + 1 + N_u - N_p\right]$,
4. $N_y = N_p + N_c$,

regardless the choice of N_p *and* N_u.

Proof Let us assume that $\boldsymbol{u}(k)$ is the optimal control vector and $\boldsymbol{u}^*(k+1)$ is the suboptimal control postulated at time $k+1$ of the form $\boldsymbol{u}^*(k+1) = [u(k+1), \ldots, u(k+N_u-1), u(k+N_u-1)]^\mathsf{T}$. Then, predictions $\bar{y}_l(k+i)$ derived at time $k+1$ are the same as these derived at time k. Therefore,

$$\bar{J}^*(k+1) = \sum_{i=2}^{N_p+1} e^2(k+i) + \rho \sum_{i=2}^{N_u} \Delta u^2(k+i-1), \tag{4.92}$$

and consequently the difference between and $\bar{J}^*(k+1)$ and $\bar{J}(k)$ can be defined as:

$$\bar{J}^*(k+1) - \bar{J}(k) = e^2(k+N_p+1) - e^2(k+1) - \rho\Delta u^2(k). \tag{4.93}$$

From terminal equality constraints (4.89) it is obvious that $e(k+N_p+1) = 0$, then

$$\bar{J}^*(k+1) - \bar{J}(k) = -e^2(k+1) - \rho\Delta u^2(k) \leq 0. \tag{4.94}$$

Let analyze the robust model beyond the prediction horizon:

$$\begin{aligned}\bar{y}_l(k+N_p+N_c) = \; & a_1\hat{y}_l(k+N_p+N_c-1)+\cdots+a_{n_a}\hat{y}_l(k+N_p+N_c-n_a)\\ & +b_1u(k+N_p+N_c-1)+\cdots+b_{n_b}u(k+N_p+N_c-n_b)\\ & +d_1u(k+N_p+N_c-1)+\cdots+d_{n_d}u(k+N_p+N_c-n_d)\end{aligned}$$

one concludes that:

(i) the external excitation of the fundamental model is constant for $N_c \geq n_b + 1 + N_u - N_p$,
(ii) as the error model is of the FIR type, its output becomes stationary after $N_c = n_d + 1 + N_u - N_p$ time steps,
(iii) taking into account the terminal equality constraints as well as the point (ii), it is clear the output of the robust model is stationary, if $N_c \geq n_a + 1$.

Consequently, tracking error equality constraints hold not only for $j \in [1, N_c]$ but for all $j \geq 1$, if

$$N_c = \max\left[n_a+1, n_b+1+N_u-N_p, n_d+1+N_u-N_p\right].$$

In the view of above deliberations, it is obvious that the predictions $\hat{y}_l$ and $\bar{y}_l$ are stationary for the horizons greater than $N_y = N_p + N_c$. An increase of the value of N_y beyond the value $N_p + N_c$ only duplicates the final constraints either on $\hat{y}_l$ or $\bar{y}_l$. Therefore, it can be concluded that the output constraints imposed on both $\hat{y}_l$ and $\bar{y}_l$ are satisfied not only for $i \in [1, N_y]$ but also for all $i \geq 1$. Finally, looking at the definition of $\boldsymbol{u}^*(k+1)$ one sees that inequality constraints (4.5) at time $k+1$ are also satisfied.

Using these considerations one can conclude that $\boldsymbol{u}^*(k+1)$ satisfies all the constraints at time $k+1$, and subsequently that the vector $\Delta\boldsymbol{u}^*(k+1)$ also satisfies the first constraint of the problem (4.91). Furthermore, if $\boldsymbol{u}(k+1)$ is the optimal solution of the optimization problem at a time $k+1$ then $\bar{J}(k+1) \leq \bar{J}^*(k+1) \leq \bar{J}(k)$ as $\boldsymbol{u}^*(k+1)$ is the suboptimal one. Finally

$$\bar{J}(k+1) - \bar{J}(k) \leq -e^2(k+1) - \rho\Delta u^2(k) \tag{4.95}$$

and it is clear that for $\rho \neq 0$ the cost is monotonically decreased with respect to time and the control system is stable.

Remark 4.4 Theorem 4.3 assumes the error model in the form of a finite impulse response system. However, much better uncertainty modelling results may be achieved using more complex systems, e.g. infinite impulse response models. In such cases however, the constraints horizon should be properly increased.

4.6 An Experimental Study — A Pneumatic Servo

In order to showcase the underlying ideas, a pneumatic servomechanism used to control the position of the mass is used as an example [29, 30]. In a pneumatic servomechanism, the transmission of signals is carried out through the medium of compressed air. Pneumatic servos are widely used in the industry because they are a low-cost solution, offer a high power-to-weight ratio, and are easy in maintenance. Moreover, they are a readily available and cheap power source. The system consists of the double-acting cylinder lifting an inertial weight. The cylinder is fed from a set of four adjustable air valves. The pneumatic servomechanism is an example of a nonlinear system. A nonlinear behaviour has, in general, the following sources: (i) nonlinear friction forces, (ii) dead-band due to stiction and (iii) dead-time due to air compressibility. In the considered plant, the friction model proposed by [15] was used:

$$F = \frac{F_s - F_c}{1 + \left(\frac{v(t)}{v_s}\right)^2}\mathrm{sign}(v(t)) + F_c\mathrm{sign}(v(t)) + k_v v(t), \tag{4.96}$$

where $v(t)$ stands for the velocity, F_s is the static friction force, v_s is the Stribeck velocity, F_c represents the Coulomb friction force, k_v is the viscous coefficient. The dead-band is modelled as follows:

$$u'(k) = \begin{cases} 0 & \text{for } -u_{lim} \geq u(k) \geq u_{lim} \\ u(k) & \text{otherwise} \end{cases}, \tag{4.97}$$

where u_{lim} is the input limiting value. Typically the dead-zone covers ca $10 \sim 12\%$ of the operational range of the servo. In this study, u_{lim} was set to 0.25.

The scheme of the process shown in Fig. 4.9, where V_1, V_2 are cylinder volumes, A_1, A_2 denotes chamber areas, P_s is the supplied pressure, P_r is the exhaust pressure, m stands for the load mass, and y is the piston position. The specification is provided in Table 4.11. More detailed information about the system can be found in [29]. The valves are operated in such a way so that for the input signal $u \geq 0$, the valves S_1 and S_4 are open. In turn, when $u < 0$ the valves S_2 and S_3 are open. All valves open proportionally to their control signals. The sampling frequency is set to 10 Hz. From the identification theory point of view, the considered pneumatic servo can be viewed as a single-input-single-output system with the control signal of valves as the input and the piston position as the output.

The pneumatic servomechanism was implemented in MATLAB®/ Simulink® software. The original simulator [29] was extended to (i) model nonlinear behaviour resulting from the Stribeck friction and dead-band, (ii) make it possible to introduce different types of uncertainty. Thus, the advantage of the simulator over the real process is that the simulator makes it possible to analyse the behaviour of the proposed predictive control in the presence of a wide range of potential uncertainty descriptions, which can be easily modelled.

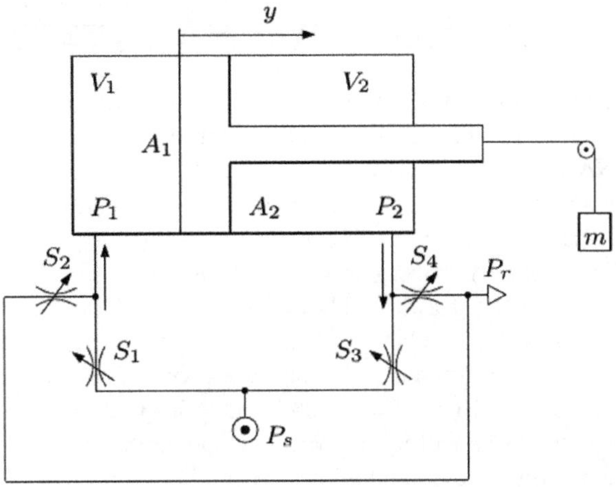

Fig. 4.9 The scheme of the pneumatic servomechanism

Table 4.11 The process specification

Symbol	Description	Value/unit	Symbol	Description	Value/unit
V_1	Cylinder 1 volume	490.8 cm^3	V_2	Cylinder 2 volume	412.3 cm^3
A_1	Chamber 1 area	19.63 cm^2	A_2	Chamber 2 area	16.49 cm^2
P_s	Supplied pressure	6 bars	P_r	Exhaust pressure	1 bar
m	Load mass	20 kg	y	Piston position	[−24.5, 24.5] cm
F_c	Coulomb friction	20 N	F_s	Static friction	30 N
v_s	Stribeck velocity	1 $\frac{\text{m}}{\text{s}}$	k_v	Viscous coefficient	30 $\frac{\text{N sec}}{\text{m}}$

4.6.1 *The Fundamental Model*

The first step of Algorithm 4.2 is to record data for neural network training. The investigated system is poorly damped and includes an integration action. For that reason, in order to collect training data it is necessary to operate in the closed-loop scheme with the classical proportional-integral (PI) controller. The reference signal should be selected in such a way as to guarantee the persistent excitation of the system. The reference was formed using a combination of harmonic signals with the frequencies from the range [0.1, 0.5] Hz, the ramp signals with different slopes and the unit steps triggered randomly with levels covering the possible piston positions from the interval (−0.245, 0.245). A reference prepared this way is applied in the closed-loop control with the PI controller described by the transfer function:

$$C(s) = K_p + \frac{1}{T_i s}. \tag{4.98}$$

The best controller settings were found to be $K_p = 30$ and $T_i = 6.7$. For the sample time $T_s = 0.1$, the discrete-time transfer function of the PI controller is in the form:

$$C(z) = \frac{30z - 29.99}{z - 1}. \tag{4.99}$$

The resulting training set consisted of 5000 samples. The recorded data are used to design a neural model of the servomechanism (Step 2 of Algorithm 4.2). The NARX network was used as the fundamental model (see Chap. 2, Sect. 2.3.1). After a number of trails, the best neural model was selected having 5 hidden neurons with hyperbolic tangent activation functions, 1 linear output neuron, $n_a = 4$ and $n_b = 2$. The training process was carried out off-line for 200 steps using the Levenberg–Marquardt algorithm. The achieved neural model was tested using another 4000 samples generated in the closed-loop control with the reference in the form of random steps. The reference trajectory is shown in Fig. 4.10. The quality of the fundamental model is $\text{SSE}(y, \hat{y}) = 0.0057$ and $\text{MSE}(y, \hat{y}) = 1.43 \cdot 10^{-6}$. The modelling results are quite satisfactory. However, as the model is used in a recursive fashion to predict the output of the plant within the prediction horizon, the quality of the 10-step ahead predictor was also investigated. Figure 4.11 shows the evolution of the $\text{SSE}(y, \hat{y})$ index along the prediction steps. As the prediction is done using one-step ahead predictor recursively it is clearly observable that the error is accumulated along the prediction steps. It should be kept in mind that the quality indexes presented in Fig. 4.11 are calculated using 4000 samples. In such a case, $\text{MSE}(y, \hat{y})$ for 10-ahead prediction is equal to $7.25 \cdot 10^{-4}$, which still can be described as being on a decent level. The quality of the model was also tested using the prediction cost J_{pred} defined as the sum of squared prediction errors calculated over the prediction horizon

$$J_{pred} = \sum_{i=1}^{N_p} (y(k+i) - \hat{y}(k+i))^2. \tag{4.100}$$

Figure 4.12b illustrates how J_{pred} changes in time. In the steady-state the prediction cost has a very low value. Large fluctuations are observed in cases when the system output changes significantly (see Fig 4.12a).

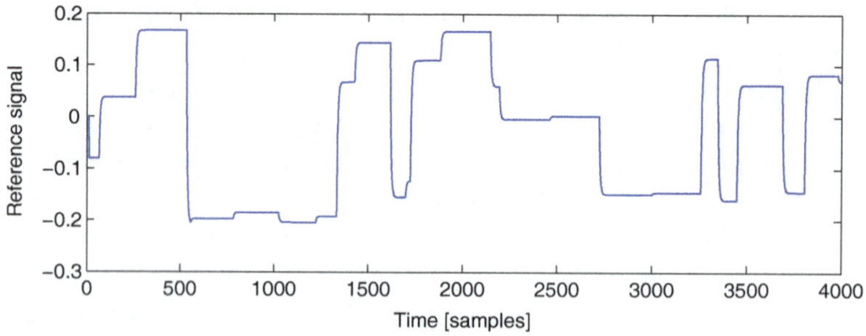

Fig. 4.10 The reference trajectory [35]. ©2018 Elsevier. Reproduced with permission

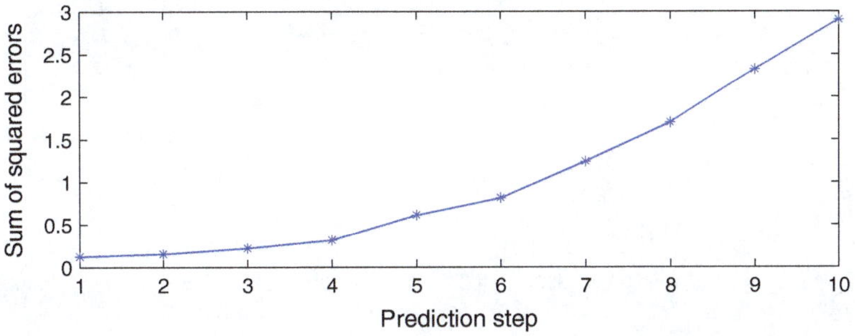

Fig. 4.11 The accumulation of the prediction error along the prediction steps [35]. ©2018 Elsevier. Reproduced with permission

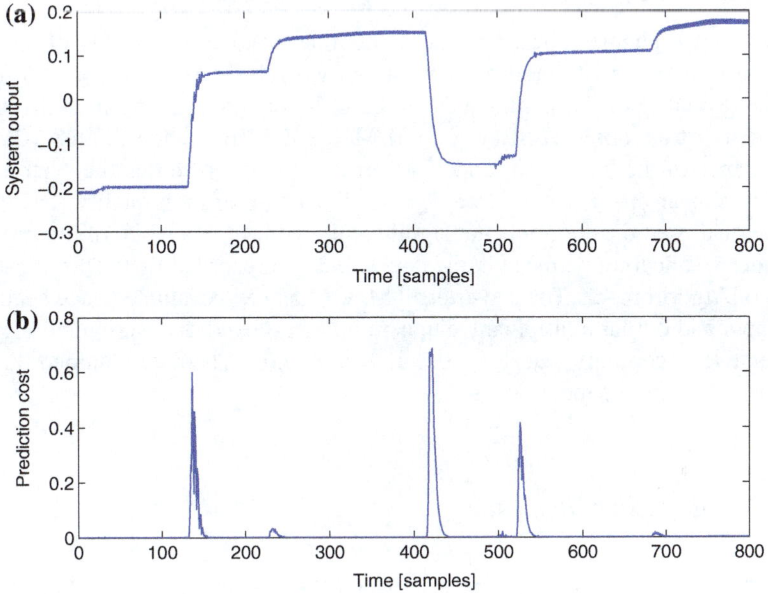

Fig. 4.12 The system output (**a**) and the prediction cost J_{pred} in time (**b**) [35]. ©2018 Elsevier. Reproduced with permission

4.6.2 Constrained MPC

The fundamental model of the plant designed in the Sect. 4.6.1 was used in the framework of constrained MPC (CMPC) (4.55) with the settings: $N_p = 10$, $N_u = 2$ and $\rho = 0.001$. Bounds imposed on variables of the investigated system are as follows: $u_{min} = -4$, $u_{max} = 4$, $y_{min} = -0.245$, $y_{max} = 0.245$. The reference trajectory is shown in Fig. 4.10. In turn, the quality of CMPC is illustrated in Fig. 4.13, where the reference is marked with dashed line and the system output by the solid line.

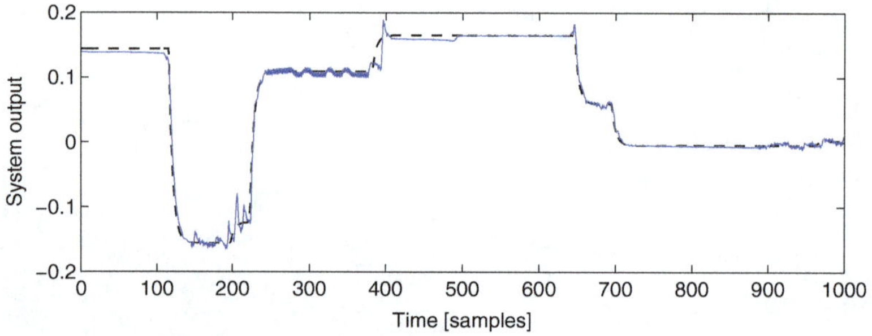

Fig. 4.13 CMPC (4.55): the reference – dashed, the system – solid [35]. ©2018 Elsevier. Reproduced with permission

As the control scheme is based on the model linearization, the quality of linearization is also investigated obtaining a satisfactory level of SSE($\hat{y}$, $\hat{y}_l$) $= 0.0572$. The modelling quality of the nonlinear model and its linearized version is also tested. The results are as follows: SSE(y, $\hat{y}$) $= 0.0386$ and SSE(y, $\hat{y}_l$) $= 0.0988$. Although performance of the linearized model is over two times worse that the performance of the nonlinear one, it can still be described as being of high quality. This means that the quality of the control is not significantly reduced as a result of linearization.

To derive uncertainty, the suitable data should be recorded first. This constitutes Step 3 of Algorithm 4.2. The constrained MPC (4.55) was evaluated once again with both input and output multiplicative uncertainty employed as described in Fig 4.14. The relative uncertainty was $\kappa_u = 0.2$ and $\kappa_y = 0.05$. The data obtained this way were used to estimate the error model.

4.6.3 Robust Performance

The performance of the proposed robust control scheme is tested using different uncertainty descriptions. The first one is the concept of multiplicative uncertainty [44] depicted in Fig. 4.14. The following two cases are investigated: uncertainty introduced at the input as well as at the output. The latter can be regarded as the

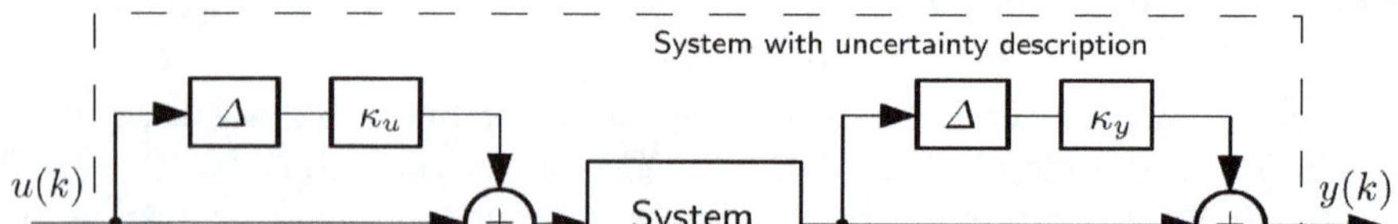

Fig. 4.14 A system with input and output multiplicative uncertainty

uncertainty of the system gain. Uncertainty observed at the input can be considered an illustration of the frequently observed fluctuations of the control signal generated by an actuator. The quantities κ_u and κ_y represent the relative uncertainty at the input $u(k)$ and at the output $y(k)$, respectively, and Δ is any real scalar satisfying $|\Delta| \leq 1$. The second kind of uncertainty is the uncertainty of the system parameters. Let $z \in [z_{min}, z_{max}]$ be a generalized system parameter, where $[z_{min}, z_{max}]$ defines the allowed space for z. The parameter z can be represented as:

$$z(k) = \bar{z}(k)(1 + \kappa_z \Delta), \qquad (4.101)$$

where $\bar{z}$ is the nominal value of z at time k, and κ_z represents the relative uncertainty in the parameter z.

It is important to emphasise that the mentioned uncertainty descriptions are introduced artificially in the considered system to test the robustness properties of the proposed control scheme. They are not used analytically during the synthesis of the control system in any way. However, looking at the design procedure (Step 3 of Algorithm 4.2), it is evident that uncertainty is taken into account implicitly through the recorded data used to identify the error model. In this way, the control system is able to deal with system uncertainty.

4.6.4 Uncertainty Modelling

Uncertainty modelling comprises Step 4 of Algorithm 4.2. The error model (4.19) in the form of both Nonlinear FIR (NFIR) and Nonlinear IIR (NIIR) model is considered. Through a series of experiments, the best models were found to be:

- NFIR model: $v_w = 4, n_c = 0, n_d = 5$, hyperbolic tangent hidden neurons, a linear output neuron.
- NIIR model: $v_w = 9, n_c = 2, n_d = 5$, hyperbolic tangent hidden neurons, a linear output neuron.

The robust model (4.21) (used in the framework of robust MPC (4.88)) is constructed using the error model. A comparison of robust models is shown in Table 4.12. Both error models guarantee that the modelling quality achieved by using the robust models (SSE$(y, \bar{y})$) is much better than the results achieved using the fundamental model (SSE$(y, \hat{y})$). The improvement is significant: 60% in the case of NFIR model and even 72% in the case of NIIR error model. The performance of the robust models was also tested using different operating regimes, namely normal operating conditions and a number of uncertainty descriptions discussed in the Sect. 4.6.3 and listed in Table 4.13. The modelling results for the NIIR error model for different operating regimes are shown in Table 4.14. In all the considered cases, the robust model is advantageous over the fundamental one. It means that robust MPC should perform better than MPC not only at the nominal operating conditions but also in case of uncertainties.

Table 4.12 A comparison of robust models [35]. ©2018 Elsevier. Reproduced with permission

Quality index	Error model	
	NFIR	NIIR
$SSE(y, \hat{y})$	0.0375	0.0387
$SSE(y, \bar{y})$	0.0228	0.0280
$\frac{SSE(y, \bar{y})}{SSE(y, \hat{y})}100\%$	60%	72%

Table 4.13 The uncertainty specification

Uncertainty/parameter	Nominal value	Relative uncertainty (%)
Multiplicative/input	–	20
Multiplicative/output	–	5
Parametric/k_v	$30\,\frac{\text{N sec}}{\text{m}}$	15
Parametric/m	20 kg	5
Parametric/V_1	$490.8\,\text{cm}^3$	18

Table 4.14 The control quality [35]. ©2018 Elsevier. Reproduced with permission

Quality index	Uncertainty in V_1	Uncertainty in m	Uncertainty in k_v	Input uncertainty	Output uncertainty
$SSE(y, \hat{y})$	0.0403	0.0588	0.046	0.0445	0.1114
$SSE(y, \bar{y})$	0.025	0.0455	0.0329	0.0292	0.0815

4.6.5 Robust MPC

The performance of the proposed robust control schemes (Step 5 of Algorithm 4.2) is tested using normal operating conditions and a number of uncertainty descriptions presented in Table 4.13. The parameters of the robust MPC as well as the reference are the same as described in the Sect. 4.6.2. Table 4.15 includes the control results represented by the index $SSE(y_r, y)$ for different operating conditions and for different control strategies: the PI controller (the second column–PI), constrained MPC (4.55) (the third column–CMPC), two variants of robust MPC (4.88) with the NFIR error model (the fourth column–RMPC-FIR) and with the NIIR error model (the fifth column–RMPC-IIR). The best indexes are in bold face. It is clear that the robust control schemes yield better result than the PI controller or even the constrained MPC. For some operating regimes, RMPC-FIR performs better than the robust scheme utilizing NIIR error model but overall RMPC-IIR is the best control scheme among the tested ones.

Table 4.15 The control quality [35]. ©2018 Elsevier. Reproduced with permission

operating regime	PI	CMPC	RMPC-FIR	RMPC-IIR
Normal operation	0.6403	0.4105	0.17	**0.1616**
Input uncertainty	0.6139	0.4304	0.6628	**0.1822**
Output uncertainty	0.6166	0.6977	**0.3748**	0.4053
Uncertainty in k_v	0.6329	0.2245	**0.17**	0.2838
Uncertainty in m	0.6174	0.6667	0.2692	**0.2208**
Uncertainty in V_1	0.6208	0.2677	0.206	**0.20**

4.6.6 Stability Considerations

To illustrate the stabilizing properties of the robust MPC, the relative uncertainty in the parameter V_1 was increased to 25%. In view of Theorem 4.1, stable robust MPC (4.91) with a FIR error model was applied. Therefore, the constraint horizon $N_c = 5$. However, in order to facilitate the optimization task, some additional actions have to be implemented. Firstly, constraint softening was applied. Equality constraints (4.89) were transformed assuming that beyond the prediction horizon the tracking error is not strictly equal to zero but its value is less that 5% of the reference amplitude:

$$-0.05y_r(k+N_p) \leq e(k+N_p+j) \leq 0.05y_r(k+N_p) \quad j \in [1, N_c]. \tag{4.102}$$

Moreover, slack variables were introduced to make the problem feasible. Secondly, before optimization began, the constraints were checked in order to remove the redundant ones. This is done by analyzing the correlation between constraints. The correlation matrix $\boldsymbol{R}$ of $\boldsymbol{X} = [\tilde{\boldsymbol{A}} \ \tilde{\boldsymbol{b}}]$ is derived:

$$\boldsymbol{R}_{i,j} = \frac{\boldsymbol{C}_{i,j}}{\sqrt{\boldsymbol{C}_{i,i}\boldsymbol{C}_{j,j}}}, \tag{4.103}$$

where $\boldsymbol{C}$ is the covariance matrix of $\boldsymbol{X}$. For each entry of $\boldsymbol{R}$ the p-value is calculated, indicating the correlation strength between the constraints. All constraints with the p-value less than 0.05 are removed. The performance of the stable robust MPC is depicted in Fig. 4.15a. It is clear that the control system preserves stability all the time even in the presence of the uncertainty in the parameters V_1. The quality of the control $\mathrm{SSE}(y_r, y) = 0.1199$. For comparison, the performance of the control system without the stability considerations is shown in Fig. 4.15b. One sees that the system has certain problems with keeping the stability between 750th and 900th time steps. The control quality is much worse as $\mathrm{SSE}(y_r, y) = 3.6932$. However, it should be kept in mind that the stable robust MPC is much more complicated from the optimization point of view, as it takes into account more constraints and the optimization might prove harder to be performed when looking for an acceptable solution. This problem is clearly visible at the beginning of the simulation, where larger fluctuations around the reference occur in the case of the stable MPC.

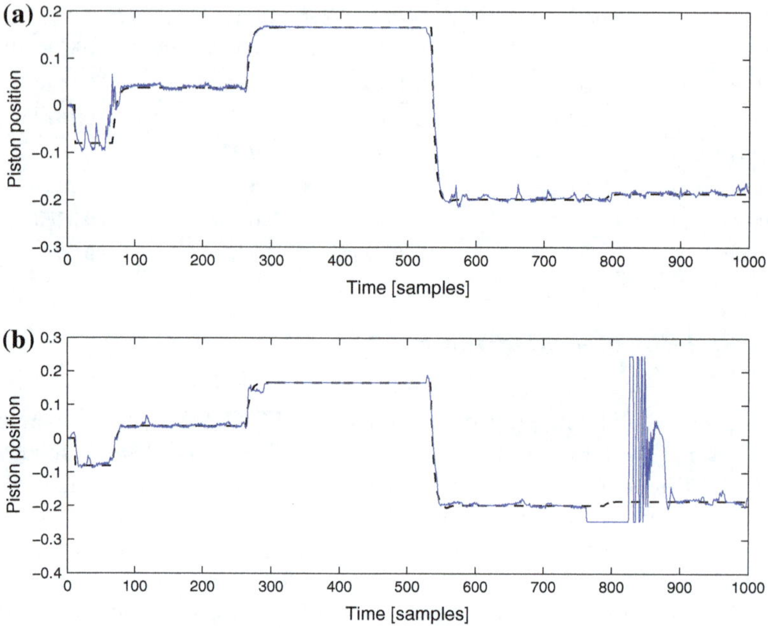

Fig. 4.15 The stability of MPC: stable robust MPC (**a**), MPC without stability considerations (**b**). Reference (dashed) and plant output (solid) [35]. ©2018 Elsevier. Reproduced with permission

4.7 Robust MPC via Statistical Bounds

One of the ways to achieve the robustness of MPC is to properly define output constraints. The cost function is the one used for the classical MPC but robustness of the control is achieved by using suitably formed output constraints, which should include the uncertainty associated with the model. The idea is to quantify the uncertainty of the neural model output in terms of the variance of the network response prediction [40] as proposed in Chap. 2, Sect. 2.5.3. This makes it possible to address the following two important issues: forming the uncertainty region around the neural network output and then reducing this uncertainty providing as reliable estimate of the system response as possible.

Let us assume that the model output is perturbed with some additive noise $w(k)$, which is customary assumed to be the realization of a zero-mean, Gaussian and white stochastic process [2, 12, 40]:

$$y(k) = \hat{y}(k) + w(k), \tag{4.104}$$

where $y(k)$ is the measured system output. In such a situation, all the measurement errors and disturbances can be represented by a component $w(k)$. This is the general way to express uncertainties for nonlinear systems described by input-output models.

The procedure for estimating the uncertainty region using optimum experimental design approach was described in detail in Chap. 2, Sect. 2.5.3 and summarized in the description of Algorithm 2.4. When the training of the neural model with the optimal data sets is finished, one is able to derive the uncertainty measure (2.99). This uncertainty measure tells us that with significance level α all possible output values are included in the region defined by a lower $w_l(k) = \hat{y}(k) - w(k)$ and an upper $w_u(k) = \hat{y}(k) + w(k)$ uncertainty estimates. This can be further used as a decision rule for defining the level of model robustness. Finally, the output constraints can be formulated as follows:

$$w_l(k+i) \leqslant \hat{y}(k+i) \leqslant w_u(k+i), \quad \forall j \in [1, N_p]. \tag{4.105}$$

However, the constraints (4.105) require one to calculate the future upper and lower estimates $w_l(k+i)$ and $w_u(k+i)$, which is troublesome, since in order to derive uncertainty (2.99), one needs to calculate the sensitivities $\frac{\partial y(\boldsymbol{u},k;\theta)}{\partial \theta}$. The calculation of these is performed on the basis of the available data. As the future inputs and output are unknown, one can only calculate the one-step-ahead prediction of uncertainty. Therefore, the constraints (4.105) should be relaxed in the following way:

$$w_l(k) \leqslant \hat{y}(k+i) \leqslant w_u(k), \quad \forall j \in [1, N_p]. \tag{4.106}$$

It means that the uncertainty measure becomes constant within the prediction horizon. Finally, the following robust MPC can be formulated:

$$\begin{aligned} \Delta \boldsymbol{u} &\triangleq \arg\min J \\ \text{s.t.} \quad \Delta u(k+N_u+j) &= 0, \quad \forall j \geq 0 \\ u_{min} \leq u(k+j) &\leq u_{max}, \quad \forall j \in [0, N_u - 1]. \\ w_l(k) \leqslant \hat{y}(k+i) &\leqslant w_u(k), \quad \forall j \in [1, N_p] \end{aligned} \tag{4.107}$$

4.8 An Experimental Study — The Pneumatic Servo Revisited

In order to preset the underlying ideas, the pneumatic servomechanism described in detail in Sect. 4.6 is investigated.

4.8.1 *Modelling*

For the reasons mentioned in Sect. 4.6, the training data was recorded in the closed-loop control with the PI controller. A section of the training data obtained this way is shown in Fig. 4.16. The entire training set consists of 16000 samples. The NARX

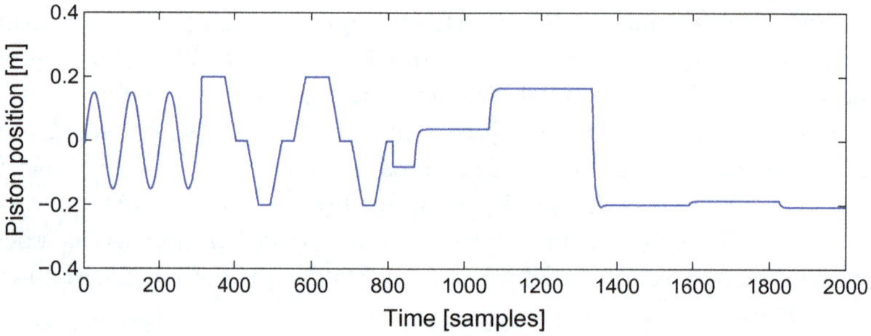

Fig. 4.16 A section of training data [38]. ©2016 IEEE. Reproduced with permission

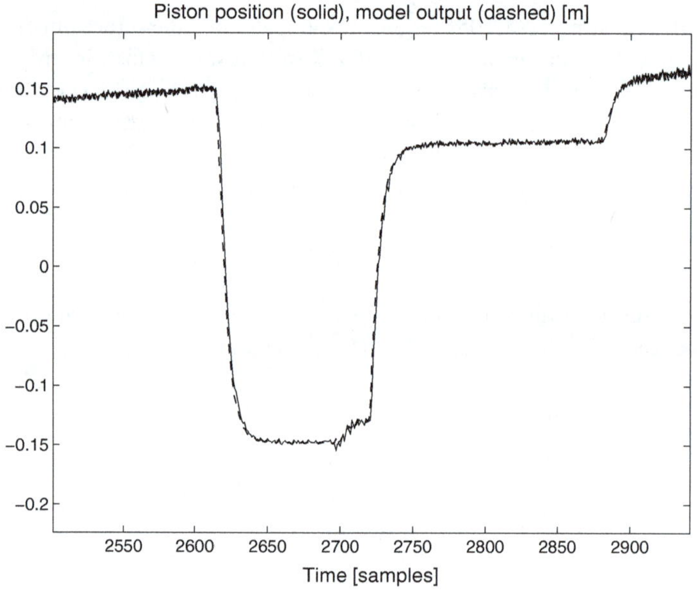

Fig. 4.17 The modelling results: the outputs of the process (solid) and the model (dashed) [38]. ©2016 IEEE. Reproduced with permission

model was used for modelling (refer to Chap. 2, Sect. 2.3.1 for details). The order of the model was selected after analyzing the physical properties of the process. The best settings for the neural network model were: $n_a = 4$, $n_b = 2$. The number of neurons $v = 5$ assures acceptable modelling results. Hidden neurons consist of the hyperbolic tangent activation function, while the output neuron has the linear activation function. The training was carried out off-line for 100 steps using the Levenberg–Marquardt algorithm. The modelling results are depicted in Fig. 4.17, where the output of the plant is marked with the solid line and the output of the model with the dashed one. For the clarity of the presentation we present only a small fragment of the results. The modelling quality is decent with $\mathrm{SSE}(y, \hat{y}) = 0.0438$ and $\mathrm{MSE}(y, \hat{y}) = 2.74 \cdot 10^{-6}$.

4.8.2 Uncertainty Modelling

A pre-trained neural network was used to select the most valuable training data. The learning data were split into 20 time sequences, containing 800 consecutive samples each. The size of the sequences was chosen arbitrarily as a trade-off between the reasonably small data package for fast training and a sufficiently informative sample properly representing the system dynamics in one sequence. The design purpose was to choose from this set of all the learning patterns the most informative sequences together with their presentation frequency. To determine the optimal design, the optimal selection problem presented in Chap. 2, Sect. 2.5.3 was applied. The goal was to choose $S = 5$ from among 20 sequences (three times reduction of dataset). All the admissible learning sequences taken with equal weights formed the initial design. The optimal design consists of the sequences 1, 3, 9, 10 and 13. This phenomenon is clearly visible in Fig. 4.18 where the optimum sequences take the maximum variance level (the crosses placed on the dashed line) and this level is significantly reduced in comparison to the maximum variance for randomly chosen data (the circles).

The objective here is to derive the output uncertainty bounds based on the results given by the optimal selection of the training sequences. According to the idea underlying Algorithm 2.4, [after the training process carried out using the optimal data sequences the network was tested with separate data set.] Setting the confidence level $1 - \alpha$ equal to 0.95, the uncertainty bounds were calculated using (2.99). The resulting uncertainty regions are illustrated in Fig. 4.19. It becomes clear that the dynamic changes of the output signal naturally increase the quality of the output prediction. As the signal becomes more stable and slow-changing, the confidence interval becomes smaller and network prediction is more reliable. At this point, using the presented methodology, one can form the uncertainty region around the output of the preliminarily trained model. Consequently, the initial model does not have to be re-trained anymore, which significantly simplifies of the model designing process.

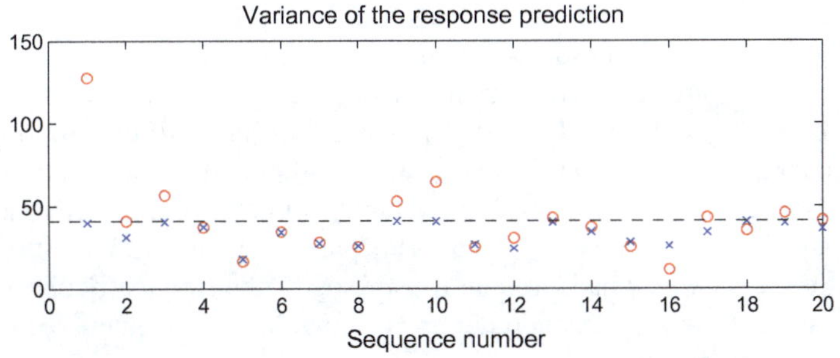

Fig. 4.18 The variances of the model response prediction for the optimum design (the crosses) and random design (the circles) [38]. ©2016 IEEE. Reproduced with permission

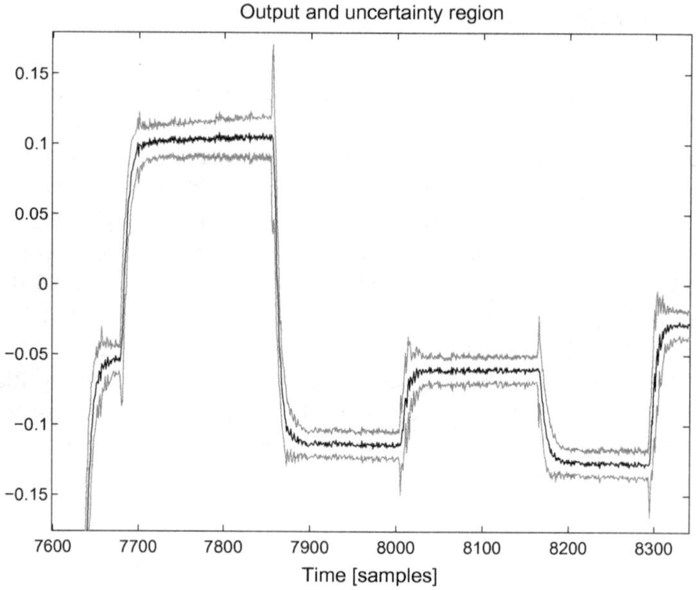

Fig. 4.19 The model output (black line) along with the uncertainty region (the grey lines) marked [38]. ©2016 IEEE. Reproduced with permission

4.8.3 *Control*

In order to design a predictive controller, the instantaneous linearization method has been applied (exhaustively discussed in Sect. 4.2.7). Such an approach facilitates the optimization by using linearization of the neural network model. At each sampling time, the neural network model is linearized. In this way, a nonlinear programming problem is converted into a quadratic programming one [8, 47], which can be easily solved using available solvers, e.g. `quadprog` in MATLAB®.

In order to achieve robustness, the output constraints including uncertainty description (4.106) are used. The parameters of predictive controller are as follows: $N_p = 10$, $N_u = 2$, $\rho = 0.002$, $k_\alpha = 0.001$. The performance of the proposed robust control schemes is tested using the concept of multiplicative uncertainty [44]. The following two cases are investigated: uncertainty introduced at the input as well as at the output. The latter can be regarded as the parametric uncertainty of the system gain. The former can be considered an illustration of the frequently observed fluctuations of the control signal generated by an actuator. Let a generalized parameter $v \in [v_{min}, v_{max}]$ be represented as:

$$v(k) = \bar{v}(k)(1 + \kappa_v \Delta), \tag{4.108}$$

Table 4.16 The control quality [38]. ©2016 IEEE. Reproduced with permission

Conditions	$SSE(y, y_r)$			δ [%]
	RMPC	MPC	PID	
Normal operation	0.3957	0.3996	1.57	1
Input uncertainty	0.349	0.4794	1.62	12.75
Gain uncertainty	0.3156	0.4058	1.59	22.23

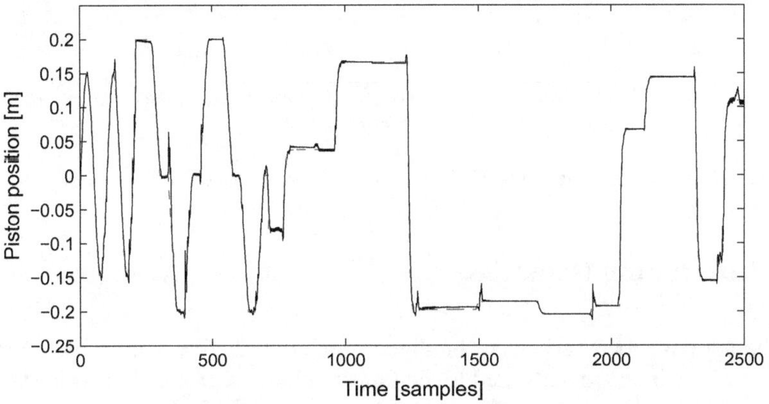

Fig. 4.20 The tracking results – RMPC [38]. ©2016 IEEE. Reproduced with permission

where $\bar{v}(k)$ is the nominal value at time k, Δ is any real scalar satisfying $|\Delta| \leqslant 1$, and κ_v represents the relative uncertainty in the parameter v. For the investigated system: $u_{min} = -4$, $u_{max} = 4$, $y_{min} = -0.245$, $y_{max} = 0.245$, relative uncertainties were set to: $\kappa_u = 0.2$, and $\kappa_y = 0.1$. Table 4.16 includes the results for MPC without robustness (the third column–MPC), robust MPC (the second column– RMPC), and the classical PI controller (the fourth column–PID). MPC uses the classical formulation of the output constraints (4.6), while RMPC uses (4.106). Table 4.16 also shows the control results in cases of different operating conditions, namely: normal operation (the first row), input multiplicative uncertainty (the second row), and output multiplicative uncertainty (the third row). In the normal operating conditions, the quality of both predictive controllers is almost the same. However, the robust controller has an advantage over the standard one (the value of the relative error δ illustrates this). Figure 4.20 shows the control results achieved by RMPC for the normal operating conditions (the reference — the dashed line, the system output — the solid line). The system output follows the reference almost immediately. One also notices a decent level of convergence of the tracking error to zero. The proposed predictive scheme is also compared with the PI controller (Fig. 4.21). It is obvious that in spite of the integral action, the PI controller has problems with tracking the reference.

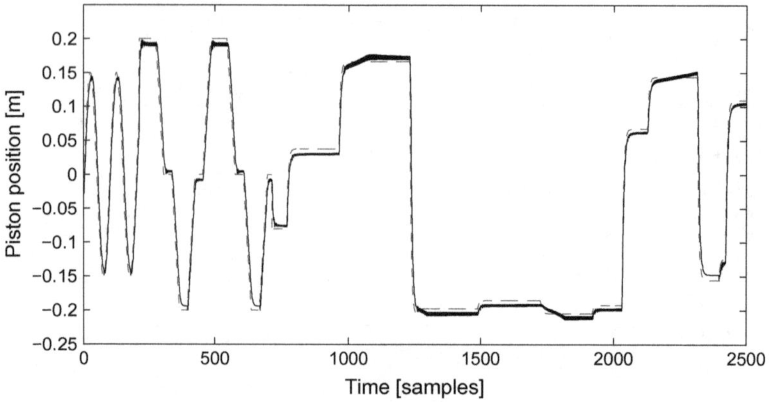

Fig. 4.21 The tracking results – PID [38]. ©2016 IEEE. Reproduced with permission

4.9 Concluding Remarks

The chapter described variants of nonlinear model predictive control based on neural networks, which are dedicated to the fault-tolerant control and the robust control schemes. The predictive controller was designed using the model of the process based on a dynamic neural network. The neural model, which is in fact a one-step-ahead predictor, can be used recursively to predict the behaviour of the plant up to N_p steps ahead. The nonlinear predictor can be directly used to determine the control sequence, which is later applied to the plant. Such a procedure leads to nonlinear constrained optimization, which can be solved using Newton-type optimization algorithms. However, one needs to bear in mind that solving nonlinear programming problems can be troublesome. Therefore in order to simplify the optimization process, the neural model can be linearized at each sampling instant. By doing this, we transform the original optimization task into the classical quadratic programming problem, which can be easily solved using numerically-efficient algorithms.

In order to build a fault-tolerant control system, MPC was equipped with a fault-diagnosis subsystem. The proposed fault-diagnosis unit was realized by means of a multi-valued binary diagnostic matrix, which was constructed on the basis of the residuals calculated using a set of partial models. Each partial model was designed in the form of a recurrent neural network. Owing to the presence of a fault-diagnosis subsystem, sensor faults can be reliably detected, isolated and finally identified, which results in estimating its size. Based on this information, the supervisory unit feeds the controller with the estimated value, which is close to the actual value. Thus, the controller ignores the changes in the measured signal and does not change control in a significant way.

Robust MPC was developed using the notion of model error modelling, which uses two neural network models: the fundamental model of the plant and the error model. The fundamental model is trained using data recorded in the closed-loop control at normal operating conditions. In turn, the error model is trained to catch the fundamental model uncertainties resulting from: (i) imperfections of the PID controller used to record the data, (ii) different kind of uncertainty observed in the plant. In order to design the error model, various neural networks were tried, i.e. NIIR and NFIR models, yielding acceptable results. The robust performance of the control system was verified by simulating multiplicative input/output and parametric uncertainties. In all the cases, the robust MPC demonstrates behaviour superior to the ones observed in the case of popular and frequently used PID controllers.

The stability of the considered control schemes was proven by checking the monotonicity of the cost function. Using the derived stability conditions, the optimization problem can be easily redefined taking into account additional constraints imposed on the process variables. The provided stability conditions guarantee the stable work of the plant even in the case of large parametric uncertainty, as illustrated by the example given.

It should be clearly stated that the quality of the neural network models strongly depends on the quality of the recorded data. Therefore, time spent on the modelling phase is of vital importance. The results presented in the chapter show that basing both fault-tolerant and robust model predictive control on neural networks should be considered a very efficient control technique.

References

1. Alessandri, A., Cervellera, C., Gaggero, M.: Predictive control of container flows in maritime intermodal terminals. IEEE Trans. Control Syst. Technol. **21**, 1423–1431 (2013)
2. Atkinson, A.C., Donev, A.N., Tobias, R.D.: Optimum Experimental Designs, with SAS. Oxford, UK (2007)
3. Bemporad, A.: A predictive controller with artificial lyapunov function for linear systems with input/state constraints. Automatica **34**(10), 1255–1260 (1998)
4. Bemporad, A., Morari, M.: Robust model predictive control: a survey. In: Garulli, A., Vicino, A., Tesi A. (eds.) Robustness in Identification and Control. Lecture Notes in Control and Information Sciences, pp. 207–226. Springer, London (1999)
5. Bertsekas, D.P.: Nonlinear Programming. Optimization and Computation Series, 2nd edn. Athena Scientific, Belmont (1999)
6. Blanke, M., Kinnaert, M., Lunze, J., Staroswiecki, M.: Diagnosis and Fault-Tolerant Control. Springer, Berlin (2006)
7. Breger, L., How, J.P.: Nonlinear model predictive control technique for unmaned air vehicles. J. Guid. Control Dyn. **29**(5), 1179–1188 (2006)
8. Camacho, E.F., Bordóns, C.: Model Predictive Control, 2nd edn. Springer, London (2007)
9. Clarke, D.W., Scattolini, R.: Constrained receding-horizon predictive control. IEE Proc. D **138**(4), 347–354 (1991)
10. Czajkowski, A., Patan, K., Szymański, M.: Application of the state space neural network to the fault tolerant control system of the PLC-controlled laboratory stand. Eng. Appl. Artif. Intell. **30**, 168–178 (2014)
11. Ducard, G.J.J.: Fault-tolerant Flight Control and Guidance Systems. Practical Methods for Small Unmanned Aerial Vehicles. Advances in Industrial Control. Springer, London (2009)
12. Fedorov, V.V., Hackl, P.: Model-Oriented Design of Experiments. Lecture Notes in Statistics. Springer, New York (1997)

13. Gossner, J.R., Kouvaritakis, B., Rossiter, J.A.: Stable generalized predictive control with constraints and bounded disturbances. Automatica **33**(4), 551–568 (1997)
14. Haykin, S.: Neural Networks. A Comprehensive Foundation, 2nd edn. Prentice-Hall, New Jersey (1999)
15. Hess, D.P., Soom, A.: Friction at a lubricated line contact operating at oscillating sliding velocities. J. Tribol. **112**, 147–152 (1990)
16. Hornik, K., Stinchcombe, M., White, H.: Multilayer feedforward networks are universal approximators. Neural Netw. **2**, 359–366 (1989)
17. Joosten, D.A., Maciejowski, J.: MPC design for fault-tolerant flight control purposes based upon an existing output feedback controller. In: Proceedings of 7th International Symposium on Fault Detection, Supervision and Safety of Technical Processes, SAFEPROCESS 2009 Barcelona, Spain, 30th June–3rd July 2009 (2009). (CD-ROM)
18. Keerthi, S.S., Gilbert, E.G.: Optimal infinite-horizon feedback laws for a general class of constrained discrete-time systems: stability and moving-horizon approximations. J. Optim. Theory Appl. **57**(2), 265–293 (1988)
19. Kim, Y.H., Kwon, W.H., Lee, Y.I.: Min-max generalized predictive control with stability. Comput. Chem. Eng. **22**(12), 1851–1858 (1998)
20. Koerber, A., King, R.: Combined feedback-feedforward control of wind turbines using state-constrained model predictive control. IEEE Trans. Control Syst. Technol. **21**, 1117–1128 (2013)
21. Korbicz, J., Kościelny, J., Kowalczuk, Z., Cholewa, W. (eds.): Fault Diagnosis. Models, Artificial Intelligence, Applications. Springer, Berlin (2004)
22. Kościelny, J.M., Bartyś, M., Syfert, M.: Method of multiple fault isolation in large scale systems. IEEE Trans. Control Syst. Technol. **20**, 1302–1310 (2012)
23. Li, Z., Xia, Y., Su, C.Y., Deng, J., Fu, J., He, W.: Missile guidance law based on robust model predictive control using neural-network optimization. IEEE Trans. Neural Netw. Learn. Syst. **26**, 1803–1092 (2015)
24. Limon, D., Bravo, J., Alamo, T., Camacho, E.F.: Robust MPC of constrained nonlinear systems based on interval arithmetic. IEE Proc. Part D. Control Theory Appl. **152**, 325–332 (2005)
25. Maciejowski, J.: Predictive Control with Constraints. Prentice-Hall, Harlow (2002)
26. Mayne, D.Q., Rawlings, J.B., Rao, C.V., Scokaert, P.O.M.: Constrained model predictive control: stability and optimality. Automatica **36**, 789–814 (2000)
27. Morari, M., Lee, J.H.: Model predictive control: past, present and future. Comput. Chem. Eng. **23**, 667–682 (1999)
28. Nelles, O.: Nonlinear System Identification. From Classical Approaches to Neural Networks and Fuzzy Models. Springer, Berlin (2001)
29. Nørgaard, M., Ravn, O., Poulsen, N., Hansen, L.: Networks for Modelling and Control of Dynamic Systems. Springer, London (2000)
30. Nørgaard, M., Sørensen, P.H., Poulsen, N., Ravn, O., Hansen, L.: Intelligent predictive control of nonlinear processes using neural networks. In: Proceedings of the 1996 IEEE International Symposium on Intelligent Control Dearborn, MI, September 15–18, 1996, pp. 301–306 (1996)
31. Noura, K., Theilliol, D., Ponsart, J.C., Chamseddine, A.: Fault Tolerant Control Systems. Design and Practical Applications. Advanced in Industrial Control. Springer, London (2009)
32. Patan, K.: Approximation of state-space trajectories by locally recurrent globally feed-forward neural networks. Neural Netw. **21**, 59–63 (2008)
33. Patan, K.: Artificial Neural Networks for the Modelling and Fault Diagnosis of Technical Processes. Lecture Notes in Control and Information Sciences. Springer, Berlin (2008)
34. Patan, K.: Neural network based model predictive control: fault tolerance and stability. IEEE Trans. Control Syst. Technol. **23**, 1147–1155 (2015)
35. Patan, K.: Two stage neural network modelling for robust model predictive control. ISA Trans. **72**, 56–65 (2018)
36. Patan, K., Korbicz, J.: Nonlinear model predictive control of a boiler unit: a fault tolerant control study. Appl. Math. Comput. Sci. **22**(1), 225–237 (2012)

37. Patan, K., Korbicz, J.: Sensor fault estimation in the framework of model predictive control. boiler case study. In: Proceedings of 8th International Symposium on Fault Detection, Supervision and Safety of Technical Processes, SAFEPROCESS 2012 Mexico City, Mexico, 2012 (2012). (CD-ROM)
38. Patan, K., Patan, M., Kowalów, D.: Optimal sensor selection for model identification in iterative learning control of spatio-temporal systems. In: 55th IEEE Conference on Decision and Control (CDC) (2016)
39. Patan, K., Witczak, P.: Robust model predictive control using neural networks. In: IEEE International Symposium on Intelligent Control, ISIC 2014, pp. 1107–1112. Antibes, France (2014)
40. Patan, M.: Sensor Networks Scheduling for Identification of Distributed Systems. Lecture Notes in Control and Information Sciences, vol. 425. Springer, Berlin (2012)
41. Rawlings, J.B., Muske, K.R.: Stability of constrained receding horizon control. IEEE Trans. Autom. Control **38**(10), 1512–1516 (1993)
42. Scokaert, P., Clarke, D.W.: Stabilizing properties of constrained predictive control. IEE Proc. Control Theory Appl. **141**(5), 295–304 (1994)
43. Scokaert, P., Mayne, D.Q., Rawlings, J.B.: Suboptimal model predictive (Feasibility implies stability). IEEE Trans. Autom. Control **44**(3), 648–654 (1999)
44. Skogestad, S., Postlethwaite, I.: Multivariable Feedback Control. Analysis and Design, 2nd edn. Wiley, New York (2005)
45. Sourander, M., Vermasvuori, M., Sauter, D., Liikala, T., Jämsä-Jounela, S.L.: Fault tolerant control for a dearomatisation process. J. Process. Control **19**, 1091–1102 (2009)
46. Staroswiecki, M., Yang, H., Jiang, B.: Active fault tolerant control based on progressive accomodation. Automatica **43**(12), 2070–2076 (2007)
47. Tatjewski, P.: Advanced Control of Industrial Processes. Springer, London (2007)
48. Tatjewski, P.: Disturbance modeling and state estimation for offset-free predictive control with state-space process models. Int. J. Appl. Math. Comput. Sci. **24**(2), 313–323 (2014)
49. Wang, L.: Model Predictive Control System Design and Implementation Using MATLAB. Springer, London (2009)
50. Yan, Z., Wang, J.: Robust model predictive control of nonlinear systems with unmodelled dynamics and bounded uncertainties based on neural networks. IEEE Trans. Neural Netw. Learn. Syst. **25**, 457–469 (2014)
51. Zangwill, W.I.: Nonlinear programming via penalty functions. Manag. Sci. **13**, 344–358 (1967)

Chapter 5
Control Reconfiguration

5.1 Introduction

Nowadays, one observes a very rapid development of the automation industry. It is not only the sophisticated and complex systems that are subject to automatisation but one also notices an increasing number of simple systems that are being fully automated. Obviously, with the growing number of plant components, the probability of system faults is also increasing. This makes it necessary to implement a system that automatically demonstrates a proper behaviour in the case of unwanted abnormal situations. Thus, the FTC systems have received increased attention in the last years [7, 9, 14]. Sensor or actuator faults, product changes, material consumption may all affect the controller performance [5, 17, 18] and can result in large economic losses or even violation of the safety regulations [1, 5].

The existing FTC methods can be divided into two groups: passive and active approaches [22]. Passive approaches are designed to work with presumed failure modes and their performance tends to be conservative, especially in the case of unanticipated faults. In contrast, active methods react to an occurrence of a system fault on-line and attempt to maintain the overall system stability and performance even in the case of unanticipated faults. This chapter describes an active approach for designing an automated fault detection and accommodation system. The detection of faults is performed on-line, and then the process of fault accommodation is carried aimed to self-correct a particular fault through a reconfiguration of the control law. The presented approach is motivated by the paper [13], where the authors realized on-line fault approximator in the form of an RBF neural network. However, instead of a neural approximator trained on-line, the neural network based state observer is proposed by us.

In spite of the fact that many effective control schemes have been described so far, including the optimal, predictive or adaptive control, many industrial installations and plants still use the standard PID controllers. In such cases, the fault-tolerant control system should be designed based on the existing control scheme with the PID

© Springer Nature Switzerland AG 2019

K. Patan, *Robust and Fault-Tolerant Control*, Studies in Systems, Decision and Control 197, https://doi.org/10.1007/978-3-030-11869-3_5

controller. The basic idea behind the proposed control reconfiguration mechanism, which is also known as fault compensation or fault accommodation, is to detect a fault first and then to recalculate the control law. In order to derive the auxiliary control, it is proposed to employ a state-space neural network model and a nonlinear observer. With such tools, it is possible to approximate a fault occurring in the plant and to recalculate the control law.

However, the auxiliary control constitutes an additional control loop which can influence the stability of the entire system. Therefore, this chapter also focuses on the issues related to the stability analysis of the proposed control scheme using the Lyapunov direct method.

5.2 Problem Formulation

Let us consider a discrete-time nonlinear dynamic system governed by the following state-space equation:

$$\boldsymbol{x}(k+1) = g(\boldsymbol{x}(k), \boldsymbol{u}(k)) + f(\boldsymbol{x}(k), \boldsymbol{u}(k)), \tag{5.1}$$

where $g(\cdot, \cdot)$ is a process working at the normal operating conditions, $\boldsymbol{x}(k)$ is the state vector, $\boldsymbol{u}(k)$ is the control and $f(\cdot, \cdot)$ represents a fault affecting a process. An unknown fault function $f(\cdot, \cdot)$ is a function of both the state and the input. Thus, it can represent a wide range of possible faults. When the process operates at the normal conditions, the fault function $f(\cdot, \cdot)$ is equal to zero. As, in general, the state vector is not fully available, in order to approximate a fault function one needs to design a model of the healthy process:

$$\bar{\boldsymbol{x}}(k+1) = \bar{g}(\bar{\boldsymbol{x}}(k), \boldsymbol{u}(k)), \tag{5.2}$$

where $\bar{\boldsymbol{x}}(k)$ is the state vector of the model working at the normal operating conditions, $\bar{g}(\cdot, \cdot)$ stands for the model of the process working at the normal operating conditions, and the state observer is represented by:

$$\hat{\boldsymbol{x}}(k+1) = \hat{g}(\hat{\boldsymbol{x}}(k), \boldsymbol{u}(k), \boldsymbol{y}(k)), \tag{5.3}$$

where $\hat{\boldsymbol{x}}(k)$ is the estimated state vector of the process, $\hat{g}(\cdot, \cdot, \cdot)$ is the mapping realized by the observer. It should be pointed out that if a function $\hat{g}(\cdot, \cdot, \cdot)$ is assumed to be unknown, then it should be approximated, e.g. by a state-space neural network performing a proper training. In this way, one achieves an approximated mapping, which is then used to control the system synthesis. Finally, $\boldsymbol{y}(k)$ is the measured output of the process. Using (5.2) and (5.3), an unknown fault function can be approximated as:

$$\hat{f}(k) = \hat{g}(\hat{\boldsymbol{x}}(k), \boldsymbol{u}(k), \boldsymbol{y}(k)) - \bar{g}(\bar{\boldsymbol{x}}(k), \boldsymbol{u}(k), \tag{5.4}$$

where $\hat{f}(k)$ is an approximation of the fault function $f(\boldsymbol{x}(k), \boldsymbol{u}(k))$. The fault effect occurring in the control system can be compensated/eliminated by correctly determining the auxiliary input $\boldsymbol{u}^{\mathtt{fa}}(k)$ based on the estimated fault function (5.4). This additional control is added to the control $\boldsymbol{u}(k)$ calculated by the standard controller. As a result, one can determine the augmented control law $\boldsymbol{u}^{\mathtt{ftc}}(k)$ as follows:

$$\boldsymbol{u}^{\mathtt{ftc}}(k) = \boldsymbol{u}(k) + \boldsymbol{u}^{\mathtt{fa}}(k). \tag{5.5}$$

The problem of determining such $\boldsymbol{u}^{\mathtt{fa}}(k)$ to compensate the effect caused by a fault can be easily solved for linear systems. Assuming that the nominal model of the system is linear

$$\boldsymbol{x}(k+1) = \boldsymbol{A}\boldsymbol{x}(k) + \boldsymbol{B}\boldsymbol{u}(k), \tag{5.6}$$

with introduction of control $\boldsymbol{u}^{\mathtt{fa}}(k)$, the process (5.1) can be rewritten in the following way:

$$\boldsymbol{x}(k+1) = \boldsymbol{A}\boldsymbol{x}(k) + \boldsymbol{B}\left(\boldsymbol{u}(k) + \boldsymbol{u}^{\mathtt{fa}}(k)\right) + f\left(\boldsymbol{x}(k), \boldsymbol{u}(k)\right), \tag{5.7}$$

where $\boldsymbol{A}$ is the state matrix and $\boldsymbol{B}$ is the control matrix. To completely compensate for the fault effect, the fault model should be as close as possible to the nominal one, therefore

$$\boldsymbol{B}\boldsymbol{u}^{\mathtt{fa}}(k) + f(\boldsymbol{x}(k), \boldsymbol{u}(k)) = 0 \tag{5.8}$$

then

$$\boldsymbol{u}^{\mathtt{fa}}(k) = -\boldsymbol{B}^{-} f\left(\boldsymbol{x}(k), \boldsymbol{u}(k)\right), \tag{5.9}$$

where $\boldsymbol{B}^{-}$ represents the pseudo-inverse of the control matrix, e.g. in the Moore-Penrose sense. Taking into account that $f(\cdot,\cdot)$ is unknown, it can be replaced with its approximation given by $\hat{f}(k)$. Finally, using (5.4) one obtains

$$\boldsymbol{u}^{\mathtt{fa}}(k) = -\boldsymbol{B}^{-}(\hat{g}(\hat{\boldsymbol{x}}(k), \boldsymbol{u}(k), \boldsymbol{y}(k)) - \bar{g}(\bar{\boldsymbol{x}}(k), \boldsymbol{u}(k))). \tag{5.10}$$

Analyzing (5.10) it is clear that in order to use the control law (5.5) and to accommodate faults in a nonlinear control system, it is required to design a nominal model of the process ($\bar{g}(\cdot,\cdot)$) and a nonlinear state observer able to catch the actual dynamics of the process ($\hat{g}(\cdot,\cdot,\cdot)$). Both models can be designed using state-space neural networks.

Moreover, in order to facilitate a derivation of the auxiliary control (5.10), the instantaneous linearization of the model in the current operating point is employed. The basic idea of fault compensation is to detect a fault first (using a fault detection block) and then, if a fault is signalled, to recalculate the control law. The block scheme of the proposed FTC scheme is presented in Fig. 5.1.

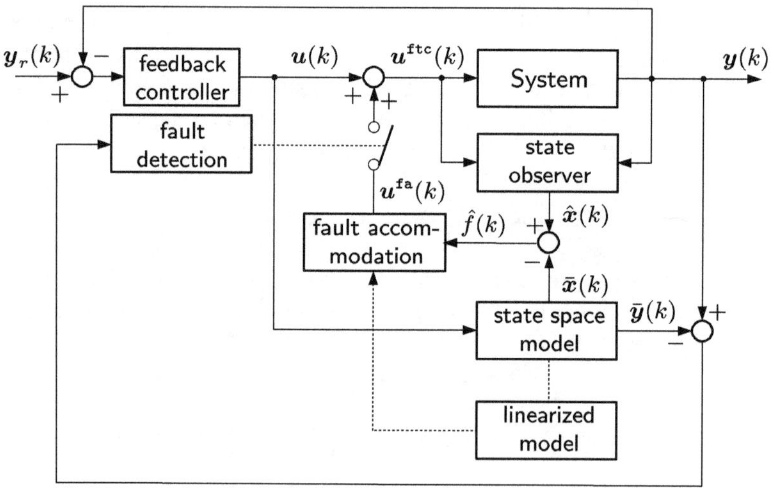

Fig. 5.1 The block scheme of the proposed control system

5.3 Process Modelling

5.3.1 A Model of the Process

In order to design a model of the considered process, a state-space neural network discussed in Chap. 2, Sect. 2.3.5 is used. The output of the hidden layer is fed back to the input layer through a bank of unit delays. The number of unit delays determines the order of the model. In general, the user decides how many neurons are used to produce feedback. Let $\boldsymbol{u}(k) \in \mathbb{R}^n$ be the input vector, $\bar{\boldsymbol{x}}(k) \in \mathbb{R}^q$ be the output of the hidden layer at time k, and $\bar{\boldsymbol{y}}(k) \in \mathbb{R}^m$ be the output vector. The state-space representation of the neural model is described by the equations:

$$\begin{aligned} \bar{\boldsymbol{x}}(k+1) &= \bar{g}(\bar{\boldsymbol{x}}(k), \boldsymbol{u}(k)) \\ \bar{\boldsymbol{y}}(k) &= \boldsymbol{C}\bar{\boldsymbol{x}}(k) \end{aligned}, \tag{5.11}$$

where $\bar{g}(\cdot, \cdot)$ is a nonlinear function characterizing the hidden layer, and $\boldsymbol{C}$ represents synaptic weights between hidden and output neurons. Introducing the weight matrix between input and the hidden layers $\boldsymbol{W}^u$ and the matrix of recurrent links $\boldsymbol{W}^x$, the representation (5.11) can be rewritten in the following form:

$$\begin{aligned} \bar{\boldsymbol{x}}(k+1) &= \boldsymbol{\sigma}(\boldsymbol{W}^x\bar{\boldsymbol{x}}(k) + \boldsymbol{W}^u\boldsymbol{u}(k)) \\ \bar{\boldsymbol{y}}(k) &= \boldsymbol{C}\bar{\boldsymbol{x}}(k) \end{aligned}, \tag{5.12}$$

where $\boldsymbol{\sigma}(\cdot)$ stands for the vector-valued activation function of the hidden neurons. In most cases, the hyperbolic tangent activation function is selected giving acceptable modelling results. For the state-space model the outputs of hidden neurons that constitute feedbacks are, in general, unknown during the training. Therefore, state-space neural models can be trained only by minimizing the simulation error. If state measurements are available, the training can be carried out much easier using the series-parallel identification scheme, similarly as in the case of the external dynamic approach (the feed-forward network with tapped delay lines). Notwithstanding this inconvenience, state-space neural models are useful in the fault-tolerant control framework as they can be used to determine the approximation of a fault effect. As a fault effect can be represented in the state space, one can handle different kinds of faults including the multiplicative and additive ones. Therefore, SSNN models do seem a very promising approach and can prove to be useful in finding a solution to fault-compensation problems. This class of nonlinear models was used in other approaches as the nominal model of the plant, see e.g. [6].

5.3.2 A Nonlinear Observer

The proposed fault-compensation scheme requires one to estimate *on-line* the state vector of the plant. Thus, there is a need to design a state observer of the considered system. This can be carried out using the state-space innovation-form observer, introduced in Chap. 2, Sect. 2.3.5. The SSIF neural model is represented as follows:

$$\begin{aligned} \hat{\boldsymbol{x}}(k+1) &= \hat{g}(\hat{\boldsymbol{x}}(k), \boldsymbol{u}(k), \varepsilon(k)) \\ \hat{\boldsymbol{y}}(k) &= \boldsymbol{C}\hat{\boldsymbol{x}}(k) \end{aligned}, \tag{5.13}$$

where $\hat{\boldsymbol{x}}(k)$ is the estimated state vector, $\varepsilon(k)$ is the error between the observer output $\hat{\boldsymbol{y}}(k)$ and the measured system output $\boldsymbol{y}(k)$. Introducing weight matrices of the neural observer, the Eq. (5.13) can be rewritten in the following form:

$$\begin{aligned} \hat{\boldsymbol{x}}(k+1) &= \boldsymbol{\sigma}(\boldsymbol{W}^x\hat{\boldsymbol{x}}(k) + \boldsymbol{W}^u\boldsymbol{u}(k) + \boldsymbol{W}^\varepsilon\varepsilon(k)) \\ \hat{\boldsymbol{y}}(k) &= \boldsymbol{C}\hat{\boldsymbol{x}}(k) \end{aligned}, \tag{5.14}$$

where $\boldsymbol{W}^x$ is the matrix of recurrent links, $\boldsymbol{W}^u$ is the input weight matrix, $\boldsymbol{W}^\varepsilon$ represents the error weight matrix, $\boldsymbol{C}$ stands for synaptic weights between the hidden and the output neurons and $\boldsymbol{\sigma}(\cdot)$ is the vector-valued activation function of the hidden neurons. In this way, using the estimated state $\hat{\boldsymbol{x}}(k)$, the unknown fault function $f(\cdot,\cdot)$ can be approximated using (5.4).

5.3.3 A Linearization of the State-Space Model

In order to simplify the fault compensation implementation, the auxiliary control is derived based on a linearized model of the plant. To achieve this, the so-called instantaneous linearization is applied. The key idea is very simple. At each sampling time a linear model is extracted from the state-space neural model. The linearization can be carried out by expanding the model into the Taylor series around the current operating point $(\boldsymbol{x}, \boldsymbol{u}) = (\boldsymbol{x}(\tau), \boldsymbol{u}(\tau))$, rejecting nonlinear components. When $\Delta\boldsymbol{x} = \bar{\boldsymbol{x}}(k) - \boldsymbol{x}(\tau)$ and $\Delta\boldsymbol{u} = \boldsymbol{u}(k) - \boldsymbol{u}(\tau)$ are taken into the account, the state-space model (5.12) expanded into the Taylor series of the first order has the following form:

$$\begin{aligned}\boldsymbol{\sigma}(\bar{\boldsymbol{x}}(k), \boldsymbol{u}(k)) &= \boldsymbol{\sigma}(\boldsymbol{x}(\tau), \boldsymbol{u}(\tau)) + \frac{\partial\boldsymbol{\sigma}}{\partial\bar{\boldsymbol{x}}}\Big|_{(\boldsymbol{x},\boldsymbol{u})}\Delta\boldsymbol{x} + \frac{\partial\boldsymbol{\sigma}}{\partial\boldsymbol{u}}\Big|_{(\boldsymbol{x},\boldsymbol{u})}\Delta\boldsymbol{u} \\ &= \boldsymbol{\sigma}(\boldsymbol{x}(\tau), \boldsymbol{u}(\tau)) + \boldsymbol{\sigma}'\boldsymbol{W}^x(\bar{\boldsymbol{x}}(k) - \boldsymbol{x}(\tau)) \\ &\quad + \boldsymbol{\sigma}'\boldsymbol{W}^u(\boldsymbol{u}(k) - \boldsymbol{u}(\tau)).\end{aligned} \tag{5.15}$$

As a result, the linear state-space model of the form:

$$\begin{aligned}\bar{\boldsymbol{x}}(k+1) &= \boldsymbol{A}\bar{\boldsymbol{x}}(k) + \boldsymbol{B}\boldsymbol{u}(k) + \boldsymbol{D} \\ \bar{\boldsymbol{y}}(k) &= \boldsymbol{C}\bar{\boldsymbol{x}}(k)\end{aligned}, \tag{5.16}$$

is obtained, where $\boldsymbol{A} = \boldsymbol{\sigma}'\boldsymbol{W}^x$, $\boldsymbol{B} = \boldsymbol{\sigma}'\boldsymbol{W}^u$, $\boldsymbol{D} = \boldsymbol{x}(\tau+1) - \boldsymbol{A}\boldsymbol{x}(\tau) - \boldsymbol{B}\boldsymbol{u}(\tau)$. The symbol $\boldsymbol{\sigma}'$ represents the first derivative of the activation function. If one uses the hyperbolic tangent as the activation function ($\sigma(z) = \tanh(z)$), this derivative can be simply calculated as:

$$\tanh(z)' = 1 - \tanh(z)^2. \tag{5.17}$$

Clearly, the outputs of the hidden neurons are used directly in order to calculate the derivatives. This is a very useful property as the linearization can be performed very quickly and thus the computation time of simulations carried out in real-time is not prolonged significantly.

5.4 Fault Detection

When the process operates in the nominal conditions, in an ideal situation, the compensation term $\boldsymbol{u}^{\mathtt{fa}}(k)$ should be equal to zero. As we cannot expect that the designed model of the plant is a faithful replica of the plant dynamics, the residual defined as a difference between the output of the plant and the output of the model is not strictly equal to zero. Thus, the fault-detection block is necessary to provide information about the current working conditions of the plant and to initiate the compensation procedure when a fault occurs. Such an approach prevents the situation where the compensation disturbs a healthy process.

Therefore, uncertainty of a model is something that should be considered during fault diagnosis. This leads to the so-called robust identification. The robust identification procedure should deliver not only a model of a given process, but also a reliable estimate of the uncertainty associated with the model [12]. One of the possible solutions is to apply the MEM procedure introduced in Chap. 2, Sect. 2.5.2.

Using a procedure analogous to the one presented in Algorithm 2.3, the centre of the uncertainty region is derived first and around it a confidence region is then formed.

The error model can be realized using different techniques. The simplest way is to use the FIR filter, as discussed in [15] but one can also use the ARX model or its nonlinear version. The pseudo-code of the proposed fault accommodation algorithm is presented in Algorithm 5.1.

Algorithm 5.1 Fault accommodation

```
Initialization: design SSNN (5.12), SSIF (5.14) and MEM (Algorithm 2.3) models;
while simulation is running do
    read process variables;
    calculate output of models (SSNN, SSIF and MEM);
    with SSNN and MEM calculate uncertainty bounds;
    if fault is detected then
        perform state-space model linearization;
        calculate and apply a new control law;
    end
    go to the next iteration of simulation
end
```

5.5 Stability Analysis

A stability analysis of the control system is extremely important in industrial applications. Lack of a stable behaviour can result in an unanticipated control signal. In the most optimistic scenario, this would lead to the stopping of the plant by emergency systems. In the worst case, it can lead to a significant damage of the components or even to a catastrophic effect on the environment, if the controlled system is of a high-risk type. In this chapter, the second method of Lyapunov is used to determine the stability conditions for the system [2, 3]. Two cases of the control system are considered: with the proportional controller (P controller) and the proportional-integral controller (PI controller).

5.5.1 P Controller

Let us assume the existence of a proportional controller. Then

$$\boldsymbol{u}(k) = K\boldsymbol{e}(k), \tag{5.18}$$

where K is a proportional term and $\boldsymbol{e}(k)$ represents a regulation error defined as:

$$\boldsymbol{e}(k) = \boldsymbol{y}_r(k) - \boldsymbol{y}(k), \tag{5.19}$$

where $\boldsymbol{y}_r(k)$ is the reference signal and $\boldsymbol{y}(k)$ is the measured system output. Using the equivalence rule [16], the regulation error can be rewritten in the form:

$$\boldsymbol{e}(k) = \boldsymbol{y}_r(k) - \bar{\boldsymbol{y}}(k) = \boldsymbol{y}_r(k) - \boldsymbol{C}\bar{\boldsymbol{x}}(k). \tag{5.20}$$

Finally, the standard control can be represented as:

$$\boldsymbol{u}(k) = K\left(\boldsymbol{y}_r(k) - \boldsymbol{C}\bar{\boldsymbol{x}}(k)\right). \tag{5.21}$$

Next, let us consider the compensation component,

$$\boldsymbol{u}^{\mathtt{fa}}(k) = -\boldsymbol{B}^{-}\hat{\boldsymbol{f}}(k) = -\boldsymbol{B}^{-}\left(\hat{\boldsymbol{x}}(k) - \bar{\boldsymbol{x}}(k)\right). \tag{5.22}$$

Substituting (5.21) and (5.22) into (5.5), one obtains

$$\boldsymbol{u}^{\mathtt{ftc}}(k) = K\left(\boldsymbol{y}_r(k) - \boldsymbol{C}\bar{\boldsymbol{x}}(k)\right) - \boldsymbol{B}^{-}\left(\hat{\boldsymbol{x}}(k) - \bar{\boldsymbol{x}}(k)\right). \tag{5.23}$$

Using (5.23), the state of the system fed with the augmented control is described as follows:

$$\bar{\boldsymbol{x}}(k+1) = \boldsymbol{\sigma}\left(\left(\boldsymbol{W}^{x} - \boldsymbol{W}^{u}K\boldsymbol{C} + \boldsymbol{W}^{u}\boldsymbol{B}^{-}\right)\bar{\boldsymbol{x}} - \boldsymbol{W}^{u}\boldsymbol{B}^{-}\hat{\boldsymbol{x}} + \boldsymbol{W}^{u}K\boldsymbol{y}_r(k)\right). \tag{5.24}$$

The state equation of the observer is:

$$\hat{\boldsymbol{x}}(k+1) = \boldsymbol{\sigma}\left(\boldsymbol{W}^{x}\hat{\boldsymbol{x}}(\boldsymbol{k}) + \boldsymbol{W}^{u}\boldsymbol{u}^{\mathtt{ftc}}(k) + \boldsymbol{W}^{e}\varepsilon(k)\right), \tag{5.25}$$

where

$$\varepsilon(k) = \hat{\boldsymbol{y}}(k) - \bar{\boldsymbol{y}}(k) = \boldsymbol{C}\left(\hat{\boldsymbol{x}}(k) - \bar{\boldsymbol{x}}(k)\right). \tag{5.26}$$

Substituting (5.23) and (5.26) into (5.25), one obtains

$$\begin{aligned}\hat{\boldsymbol{x}}(k+1) = \boldsymbol{\sigma}\big(& \left(-\boldsymbol{W}^{e}\boldsymbol{C} - \boldsymbol{W}^{u}K\boldsymbol{C} + \boldsymbol{W}^{u}\boldsymbol{B}^{-}\right)\bar{\boldsymbol{x}}(k) \\ & + \left(\boldsymbol{W}^{x} - \boldsymbol{W}^{u}\boldsymbol{B}^{-} + \boldsymbol{W}^{e}\boldsymbol{C}\right)\hat{\boldsymbol{x}}(k) + \boldsymbol{W}^{u}K\boldsymbol{y}_r(k)\big).\end{aligned} \tag{5.27}$$

Introducing the augmented state vector $\boldsymbol{x}(k) = [\bar{\boldsymbol{x}}^{\mathsf{T}}(k)\,\hat{\boldsymbol{x}}^{\mathsf{T}}(k)]^{\mathsf{T}}$, the state equation of the control system can be represented as follows:

$$\boldsymbol{x}(k+1) = \boldsymbol{\sigma}\left(\mathcal{A}\boldsymbol{x}(k) + \mathcal{B}\right), \tag{5.28}$$

where

$$\mathcal{A} = \begin{bmatrix} \boldsymbol{W}^{x} - \boldsymbol{W}^{u}K\boldsymbol{C} + \boldsymbol{W}^{u}\boldsymbol{B}^{-} & -\boldsymbol{W}^{u}\boldsymbol{B}^{-} \\ -\boldsymbol{W}^{e}\boldsymbol{C} - \boldsymbol{W}^{u}K\boldsymbol{C} + \boldsymbol{W}^{u}\boldsymbol{B}^{-} & \boldsymbol{W}^{x} - \boldsymbol{W}^{u}\boldsymbol{B}^{-} + \boldsymbol{W}^{e}\boldsymbol{C} \end{bmatrix} \tag{5.29}$$

and

$$\mathcal{B} = \begin{bmatrix} \boldsymbol{W}^u K \boldsymbol{y}_r(k) \\ \boldsymbol{W}^u K \boldsymbol{y}_r(k) \end{bmatrix}. \tag{5.30}$$

5.5.2 *PI Controller*

Let us assume a discrete state-space representation of the PI controller:

$$\begin{cases} \boldsymbol{x}_r(k+1) = \boldsymbol{x}_r(k) + K_i \boldsymbol{e}(k) \\ \boldsymbol{u}(k) = \boldsymbol{x}_r(k) + (K_p + K_i)\boldsymbol{e}(k) \end{cases}, \tag{5.31}$$

where K_p is the proportional term, K_i is the integral term. Using (5.20) the standard control can be represented as:

$$\begin{cases} \boldsymbol{x}_r(k+1) = \boldsymbol{x}_r(k) + K_i \left(\boldsymbol{y}_r(k) - \boldsymbol{C}\bar{\boldsymbol{x}}(k)\right) \\ \boldsymbol{u}(k) = \boldsymbol{x}_r(k) + (K_p + K_i)\left(\boldsymbol{y}_r(k) - \boldsymbol{C}\bar{\boldsymbol{x}}(k)\right) \end{cases}. \tag{5.32}$$

A redefined PI controller with applied fault-compensation mechanism can be rewritten in the form:

$$\begin{cases} \boldsymbol{x}_r(k+1) = \boldsymbol{x}_r(k) + K_i \boldsymbol{e}(k) \\ \boldsymbol{u}^{\mathtt{ftc}}(k) = \boldsymbol{x}_r(k) + (K_p + K_i)\boldsymbol{e}(k) + \boldsymbol{u}^{\mathtt{fa}}(k) \end{cases}, \tag{5.33}$$

where $\boldsymbol{u}^{\mathtt{fa}}(k)$ is auxiliary/compensating control value represented by (5.22). Substituting (5.32) and (5.22) in (5.5), one obtains the following rule:

$$\boldsymbol{u}^{\mathtt{ftc}}(k) = \boldsymbol{x}_r(k) + (K_p + K_i)\left(\boldsymbol{y}_r(k) - \boldsymbol{C}\bar{\boldsymbol{x}}(k)\right) - \boldsymbol{B}^-\left(\hat{\boldsymbol{x}}(k) - \bar{\boldsymbol{x}}(k)\right). \tag{5.34}$$

Applying the augmented control (5.34), the state of the system (5.12) becomes

$$\begin{aligned} \bar{\boldsymbol{x}}(k+1) = \boldsymbol{\sigma}\big(&\left(\boldsymbol{W}^x - (K_p + K_i)\boldsymbol{W}^u\boldsymbol{C} + \boldsymbol{W}^u\boldsymbol{B}^-\right)\bar{\boldsymbol{x}}(k) \\ &- \boldsymbol{W}^u\boldsymbol{B}^-\hat{\boldsymbol{x}}(k) + \boldsymbol{W}^u\boldsymbol{x}_r(k) + (K_p + K_i)\boldsymbol{W}^u\boldsymbol{y}_r(k)\big). \end{aligned} \tag{5.35}$$

Substituting (5.34) and (5.26) into the state equation of the observer (5.25), one obtains

$$\begin{aligned} \hat{\boldsymbol{x}}(k+1) = \boldsymbol{\sigma}\big(&\left(-\boldsymbol{W}^e\boldsymbol{C} - (K_p + K_i)\boldsymbol{W}^u\boldsymbol{C} + \boldsymbol{W}^u\boldsymbol{B}^-\right)\bar{\boldsymbol{x}}(k) \\ &+ \left(\boldsymbol{W}^x - \boldsymbol{W}^u\boldsymbol{B}^- + \boldsymbol{W}^e\boldsymbol{C}\right)\hat{\boldsymbol{x}}(k) \\ &+ \boldsymbol{W}^u\boldsymbol{x}_r(k) + (K_p + K_i)\boldsymbol{W}^u\boldsymbol{y}_r(k)\big). \end{aligned} \tag{5.36}$$

Applying the augmented state $\boldsymbol{x}(k) = [\bar{\boldsymbol{x}}^{\mathsf{T}}(k)\ \hat{\boldsymbol{x}}^{\mathsf{T}}(k)\ \boldsymbol{x}_r^{\mathsf{T}}(k)]^{\mathsf{T}}$, the state equation of the control system can be represented as follows:

$$\boldsymbol{x}(k+1) = \boldsymbol{\sigma}\left(\mathcal{A}\boldsymbol{x}(k) + \mathcal{B}\right), \tag{5.37}$$

where

$$\mathcal{A} = \begin{bmatrix} \boldsymbol{W}^x - (K_p + K_i)\boldsymbol{W}^u\boldsymbol{C} + \boldsymbol{W}^u\boldsymbol{B}^- & -\boldsymbol{W}^u\boldsymbol{B}^- & \boldsymbol{W}^u \\ -\boldsymbol{W}^e\boldsymbol{C} - (K_p + K_i)\boldsymbol{W}^u\boldsymbol{C} + \boldsymbol{W}^u\boldsymbol{B}^- & \boldsymbol{W}^x - \boldsymbol{W}^u\boldsymbol{B}^- + \boldsymbol{W}^e\boldsymbol{C} & \boldsymbol{W}^u \\ -K_i\boldsymbol{C} & 0 & 1 \end{bmatrix} \tag{5.38}$$

and

$$\mathcal{B} = \begin{bmatrix} (K_p + K_i)\boldsymbol{W}^u\boldsymbol{y}_r(k) \\ (K_p + K_i)\boldsymbol{W}^u\boldsymbol{y}_r(k) \\ K_i\boldsymbol{y}_r(k) \end{bmatrix}. \tag{5.39}$$

5.5.3 Sufficient Conditions for Stability

The general form of a control system (5.37) is the same for both types of the discussed controllers (P and PI), therefore the final part of stability analysis can be treated in a unified way. In order to apply the Lyapunov method to the system (5.37), a number of transformations are required to be performed. First, let us introduce a linear transformation in the form:

$$\boldsymbol{v}(k) = \mathcal{A}\boldsymbol{x}(k) + \mathcal{B}, \tag{5.40}$$

now, (5.37) can be rewritten as follows:

$$\boldsymbol{v}(k+1) = \mathcal{A}\boldsymbol{\sigma}\left(\boldsymbol{v}(k)\right) + \mathcal{B}. \tag{5.41}$$

Next, let introduce the equivalent coordinate transformation:

$$\boldsymbol{z}(k) = \boldsymbol{v}(k) - \boldsymbol{v}^*(k), \tag{5.42}$$

where $\boldsymbol{v}^*(k)$ is the equilibrium point of system (5.41), and assuming $\mathcal{B}$ as a threshold or a fixed point, the system (5.41) can be transformed into:

$$\boldsymbol{z}(k+1) = \mathcal{A}\bar{\boldsymbol{\sigma}}\left(\boldsymbol{z}(k)\right), \tag{5.43}$$

where $\bar{\boldsymbol{\sigma}}(\boldsymbol{z}(k)) = \boldsymbol{\sigma}\left(\boldsymbol{z}(k) + \boldsymbol{v}^*(k)\right) - \boldsymbol{\sigma}\left(\boldsymbol{v}^*(k)\right)$.

Now, one can formulate the stability conditions for the proposed control system.

Theorem 5.1 ([4]) *The control system (5.43) is globally asymptotically stable, if there exists a matrix* $\boldsymbol{P} \succ 0$*, such that the following condition is satisfied:*

$$\mathcal{A}^{\mathsf{T}}\boldsymbol{P}\mathcal{A} - \boldsymbol{P} \prec 0. \tag{5.44}$$

Proof Let us consider a positive definite candidate Lyapunov function:

$$V(z) = z^{\mathsf{T}}\boldsymbol{P}z. \tag{5.45}$$

According to the direct Lyapunov method, the difference along the trajectory of the system (5.43) is given as follows:

$$\begin{aligned}\Delta V(z(k)) &= V(z(k+1)) - V(z(k)) \\ &= (\mathcal{A}\bar{\sigma}(z(k)))^{\mathsf{T}}\boldsymbol{P}\mathcal{A}\bar{\sigma}(z(k)) - z^{\mathsf{T}}(k)\boldsymbol{P}z(k) \qquad (5.46)\\ &= \bar{\sigma}(z(k))^{\mathsf{T}}\mathcal{A}^{\mathsf{T}}\boldsymbol{P}\mathcal{A}\bar{\sigma}(z(k)) - z^{\mathsf{T}}(k)\boldsymbol{P}z(k). \qquad (5.47)\end{aligned}$$

Taking into account the following property of the activation function [10]:

$$|\bar{\sigma}(z(k))| \leq |z(k)|, \tag{5.48}$$

Equation (5.46) takes the form:

$$\begin{aligned}\Delta V(z(k)) &\leq z^{\mathsf{T}}(k)\mathcal{A}^{\mathsf{T}}\boldsymbol{P}\mathcal{A}z(k) - z^{\mathsf{T}}(k)\boldsymbol{P}z(k) \\ &\leq z^{\mathsf{T}}(k)(\mathcal{A}^{\mathsf{T}}\boldsymbol{P}\mathcal{A} - \boldsymbol{P})z(k). \qquad (5.49)\end{aligned}$$

From (5.49) it is evident that if

$$\mathcal{A}^{\mathsf{T}}\boldsymbol{P}\mathcal{A} - \boldsymbol{P} \prec 0, \tag{5.50}$$

then $\Delta V(z(k))$ is negative definite and the system (5.43) is globally asymptotically stable. □

Remark 5.2 From a practical point of view, selecting a proper matrix $\boldsymbol{P}$ in order to satisfy the condition (5.50), can be troublesome. Fortunately, the methods that use linear matrix inequalities are becoming increasingly popular among the control theory researchers owing to the simplicity and effectiveness of these methods as well as relative low numerical complexity of finding $\boldsymbol{P}$.

Assuming the above methodology, when $\boldsymbol{P}$ satisfying the condition (5.44) is found, the stable fault accommodation can be achieved. The stable fault-compensation algorithm is represented in Algorithm 5.2.

Algorithm 5.2 Stable fault compensation

```
while simulation is running do
    read process variables;
    calculate output of models (SSNN,SSIF and MEM);
    with SSNN and MEM calculate uncertainties bands;
    if fault is detected then
        perform state-space model linearization;
        calculate a new control law;
        derive matrices describing control system;
        solve LMI (5.44);
        if P satisfying the condition (5.44) is found then
            apply a new control law
        end
    end
    go to next iteration of simulation
end
```

5.6 An Experimental Study — The Tank Unit Revisited

The first investigated object is the tank unit introduced in Chap. 4. This simulation model makes it possible to generate a number of faulty situations. A specification of the faults considered is included in Table 5.1. The faults are of different nature. As one can see in Table 5.1, there are multiplicative as well as additive faults. It is also possible to set the intensity of a fault.

5.6.1 Process Modelling

Process modelling is the first step in fault-tolerant control design. In order to build a proper model, training data describing the process under normal operating conditions are required. The input signal should be as informative as possible. In the case of the tank unit, the training data were collected in open-loop control, where the input signal in the form of random steps with levels from the interval (0, 100) was used. Each step lasted 240 s. The obtained data were analyzed using a discrete Fourier transform (DFT). The DFT spectral distribution of the analyzed input including 100,000 samples is presented in Fig. 5.2. The cut-off frequency was set to $f_c =$

Table 5.1 The specification of the faulty scenarios considered

Fault	Description	Type
f_1	Fluid choking	Partly closed (0.5)
f_2	Level transducer failure	Additive (−0.05)
f_3	Positioner failure	Multiplicative (0.7)
f_4	Valve head/servo-motor fault	Multiplicative (0.8)

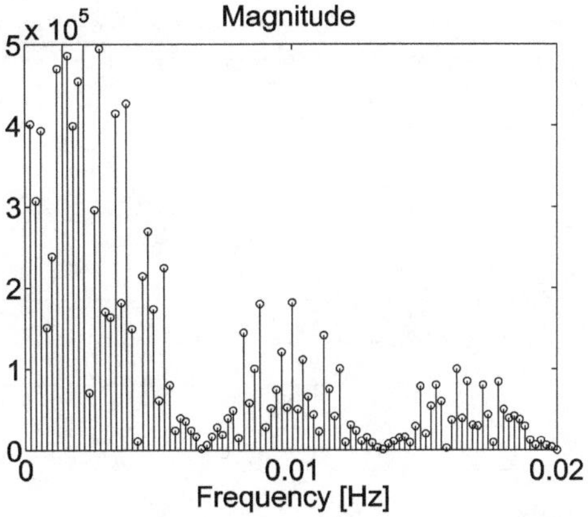

Fig. 5.2 The DFT spectrum of random steps (100,000 samples)

0.0067 Hz. This means that the data can be sampled with the period equal to 75 s. The tank unit is a process with slow dynamics. Filling the tank to the level equal to 0.25 m using a maximal input flow takes approximately 300 s. On this basis, one can assume that the distinct frequencies are lower than 0.02 Hz (i.e. the components with the period greater than 50 s). However, during the selection of the sampling time, one should pay attention to the reaction time of the fault detection block. From this point of view, the sampling period should be relatively short, in order to facilitate a fast detection of faults. Thus, the sampling period was set to 5 s as a relatively good middle value.

For the training purposes, a training set based on the data generated from the simulator of the tank unit, implemented in Simulink®, was created. The set was generated using the control signal in the form of random steps with the values from the interval $[0, 0.5]$ and consists of 500 samples. The neural network state-space innovation form (5.13) was trained for 100 epochs using the dynamic version of the LM algorithm. The model input was the control value (CV) and the model output was the level in the tank (L). Various different structures of the neural model were tried by changing the number of the hidden neurons as well as the order of the model. The best model was selected using the sum-of-squared-errors (SSE) index and the final prediction error (FPE) information criterion. The results of model development are presented in Table 5.2. Each network configuration was trained 10 times, and quality indexes were averaged.

The best results (surrounded by frames) were achieved for the second-order neural model consisting of seven hidden neurons with the hyperbolic tangent activation function. For this model, both quality indexes have the lowest values. The quality of modelling for the best neural model is presented in Fig. 5.3a. In turn, the testing of the best model is shown in Fig. 5.3b. In both cases, one can see that the output of the model follows the process output almost immediately, which testifies to a decent level of generalization capabilities of the model.

Table 5.2 The selection of the neural network structure

Network structure	Number of neurons	Model order	SSE	FPE
1	4	1	12.4918	0.000628
2	4	2	45.0943	0.0023
3	4	3	32.3772	0.0016
4	7	1	35.2229	0.0018
5	7	2	10.6031	0.000533
6	7	3	32.5675	0.0016
7	15	1	25.4094	0.0013
8	15	2	13.0330	0.000656
9	15	3	15.7853	0.000794

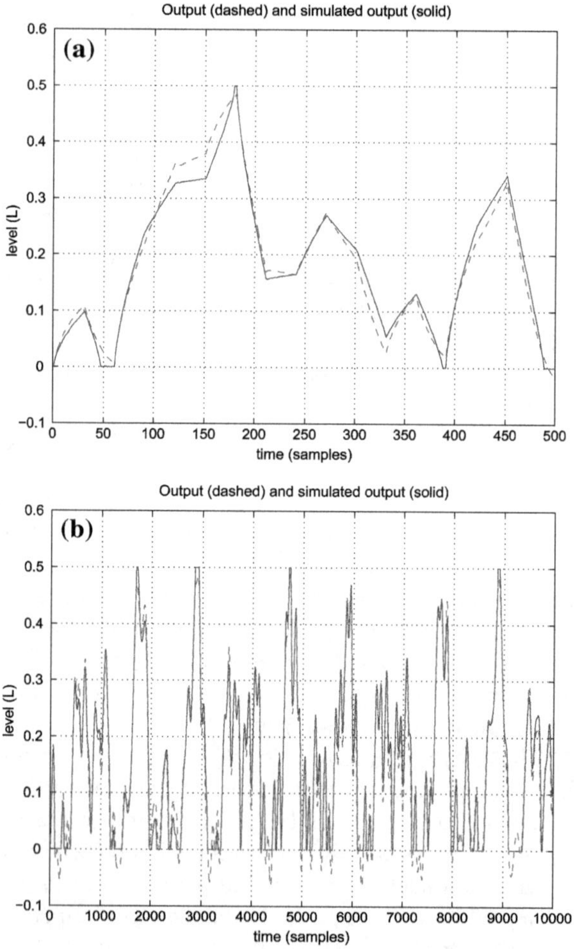

Fig. 5.3 The modelling results (process (dashed), model (solid)): training phase (**a**), testing phase (**b**)

5.6.2 Model Error Modelling

In order to obtain a more reliable method for threshold adaptation, one should estimate model uncertainty taking into consideration other process variables, e.g. the measurable process inputs and outputs. A robust model of the system consists of the state-space model developed in Sect. 5.6.1 and an error model. In order to build the error model, three different auto-regressive models were tried: the classical ARX model, a nonlinear ARX (NLARX) based on a wavelet neural network [21] and a neural-network ARX (NNARX) based on a multilayer perceptron [8]. Various different combinations of the input (n_a) and the output (n_b) delays were tested and the quality of the obtained models is listed in Table 5.3.

All models were compared taking into account the SSE index with the best result (surrounded by a frame). This best model was linear ARX with $n_a = 15$ and $n_b = 5$ with the nonlinear ARX models based on a neural network achieving worse results; however, NLARX is only marginally worse than the linear ARX. To definitely judge the quality of NLARX, further investigations should be carried out checking, for example, how the type of wavelet used influences the quality of modelling. Figure 5.4 illustrates the idea of an uncertainty region. The output of the system is marked with the solid line, while the centre of the uncertainty region with the dotted one. Using a specified significance level, confidence bands (marked with the dashed lines) are generated around the centre. The presented decision making method was evaluated taking into account two quality indexes: the time of the fault detection t_{dt} and the number of the false alarms r_{fd} [10, 11]. The results are listed in Table 5.4 for different

Table 5.3 The results of error modelling

Error model	Delays		SSE
	n_a	n_b	
ARX	1	5	13.333
	5	1	5.763
	5	15	6.3187
	15	5	**5.5946** (framed)
	15	15	6.256
NLARX	1	5	45.205
	5	1	20.408
	5	15	6.1227
	15	5	6.3858
	15	15	6.381
NNARX	1	5	15441
	5	1	42.632
	5	15	53.005
	15	5	206.2735
	15	15	50.488

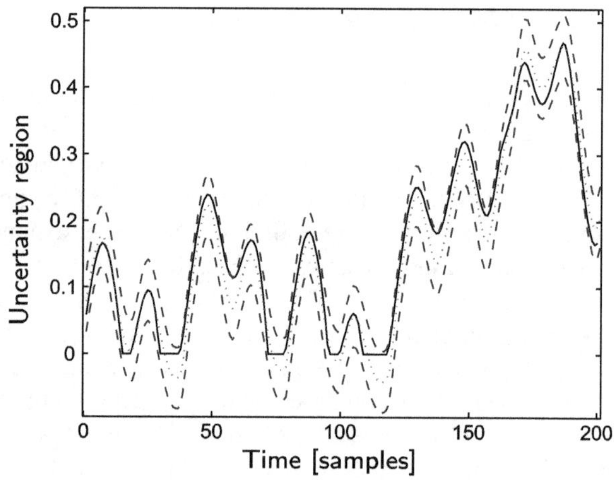

Fig. 5.4 An illustration of MEM decision making

Table 5.4 The quality indexes of the investigated decision making method

$1-\alpha$	85%	98%	99.9%
t_{dt}	10	20	25
r_{fd}	0.1339	0.0117	0.0039

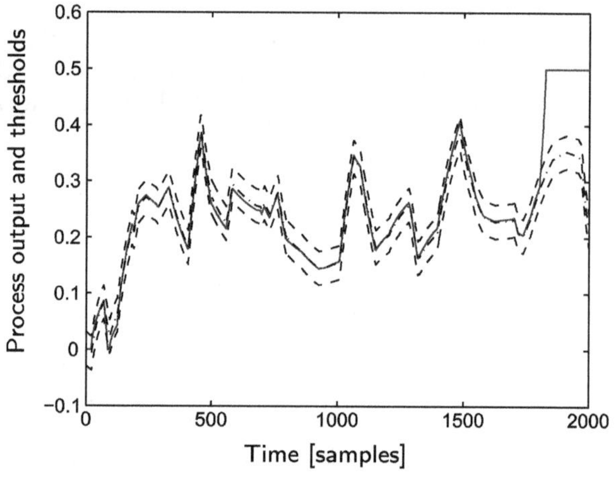

Fig. 5.5 Model error modelling: the system output (solid) and the uncertainty bounds (dashed)

values of the confidence level. As expected, by increasing the value of the confidence level, the number of false alarms decreases and, at the same time, the time of fault detection increases. It means that the fault detection system needs more time to point out abnormal operating conditions of the plant. Taking into account the value of the sampling time, which is equal to 5 s, the system can signal the fault relatively quickly; in the best case, it needs two sampling periods to do that. The fault detection abilities of MEM are shown in Fig. 5.5. In this scenario, a fault was simulated at the 1800th time instant. One can see that the method of model error modelling exhibits a very reliable behaviour, since the fault is clearly observed and quickly detected.

5.6.3 *Fault Compensation*

In order to check the fault-accommodation approach proposed in this work, a number of experiments were carried out investigating the behaviour of the system control in the cases of various faults that were introduced. The objective of the control system working with the PI controller was to keep the constant level in the tank equal to 0.25 m. Each fault listed in Table 5.1 was simulated at the 500th time instant. In the nominal conditions, the value of the additional control $\boldsymbol{u}^{\mathrm{fa}}$ is set to 0, but in the case of a fault its value is changing according to (5.10); then the fault effect is compensated. Figure 5.6a–d present the behaviour of the fault-tolerant system in the case of faults introduced at the 500th time instant, listed in Table 5.1. Each chart presents the output of the healthy system (the solid line), the output of the faulty system without compensation (the dash-dot line), and the outputs of the compensated system using model error modelling (the dashed line). In turn, the fault-tolerant control results are given in Table 5.5. All faults were quickly compensated apart from the fault f_2. In this case, the system without compensation (using only the PI controller) works better. In any case, the fault-accommodation scheme works on a decent level and the fault effect is quickly compensated, which means that the behaviour of the fault-tolerant control can be significantly increased.

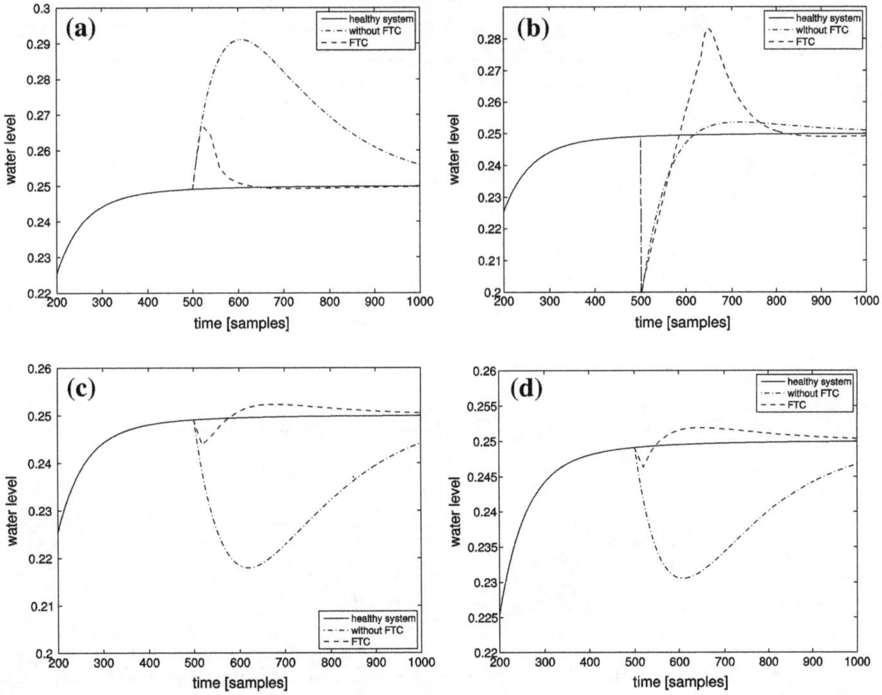

Fig. 5.6 A comparison of PI and FTC control with different fault detection methods in the case of f_1 (**a**), f_2 (**b**), f_3 (**c**) and f_4 (**d**) fault scenarios

Table 5.5 The fault-tolerance quality measures

	Model error modelling			
	f_1	f_2	f_3	f_4
SSE^{noftc}	0.3448	0.0626	0.2188	0.0756
SSE^{ftc}	0.0134	0.1302	0.0023	0.0012

5.6.4 Stability Analysis

The final part of the experiments with the tank simulator concerned the stability analysis of the proposed FTC scheme. The evaluation of the presented method was carried out in two ways. First, it was verified experimentally through visual inspection of the system states, and numerically through finding the matrix $\boldsymbol{P}$ using the LMI technique in reference to Theorem 5.1. The convergence of system states of the original system (5.37) and its autonomous version (5.43) is considered. The nominal states with no faults are presented in Fig. 5.7a for the original states and in Fig. 5.7b for the transformed autonomous system. The results for the original states achieved in the case of the evaluated faults are shown in Figs. 5.8a, 5.9a, 5.10a and 5.11a. Figures 5.8b, 5.9b, 5.10b and 5.11b presented the states of the transformed autonomous system. All the states converge to zero, which means that the control system is stable.

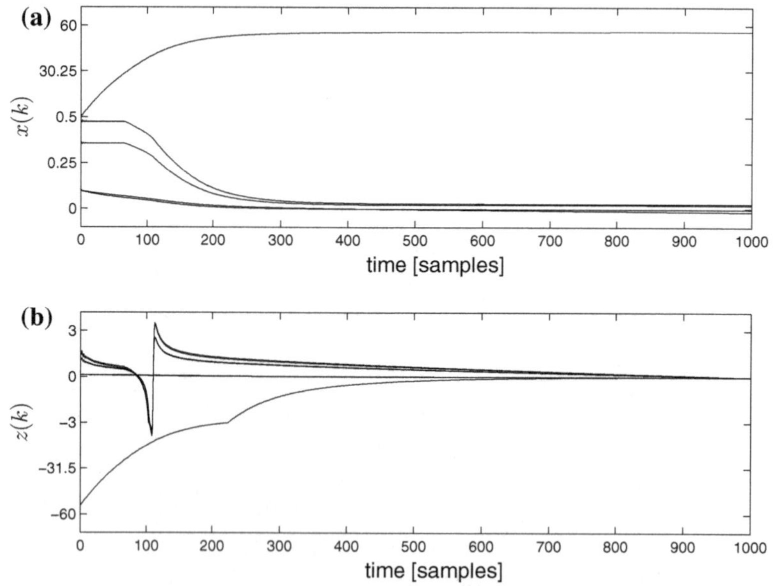

Fig. 5.7 The convergence of the control-system states without a fault: the original system (**a**), the transformed system (**b**)

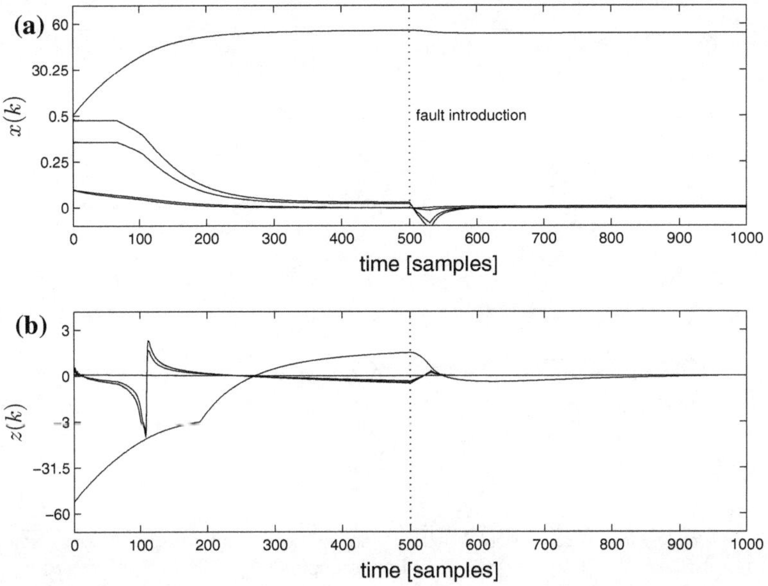

Fig. 5.8 The convergence of the control-system states with a faulty (f_1) system: the original system (**a**), the transformed system (**b**)

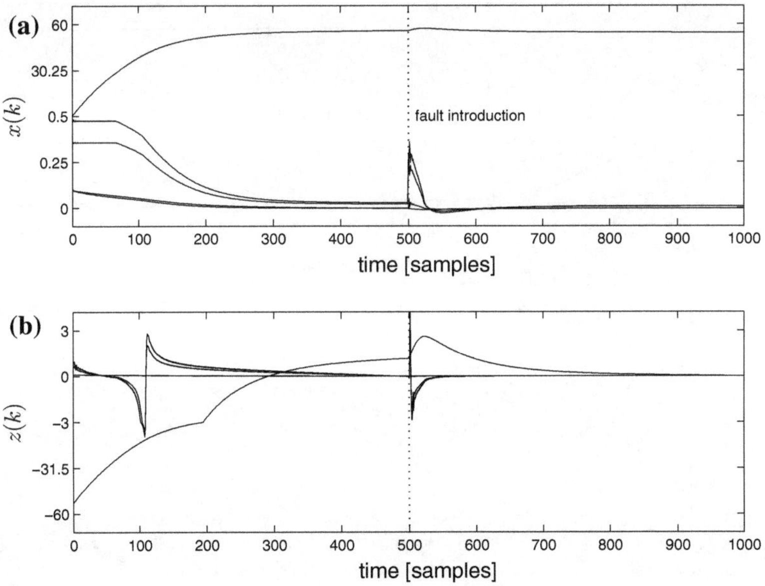

Fig. 5.9 The convergence of the control-system states with a faulty (f_2) system: the original system (**a**), the transformed system (**b**)

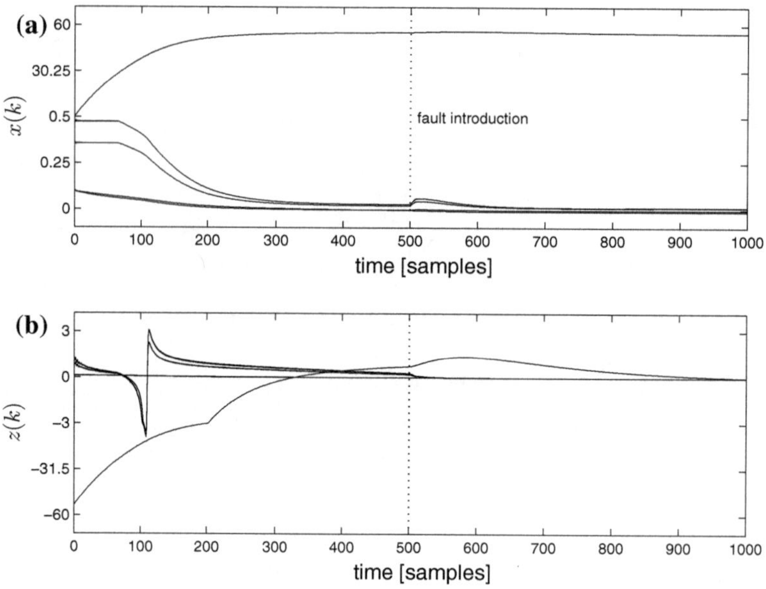

Fig. 5.10 The convergence of the control-system states with a faulty (f_3) system: the original system (**a**), the transformed system (**b**)

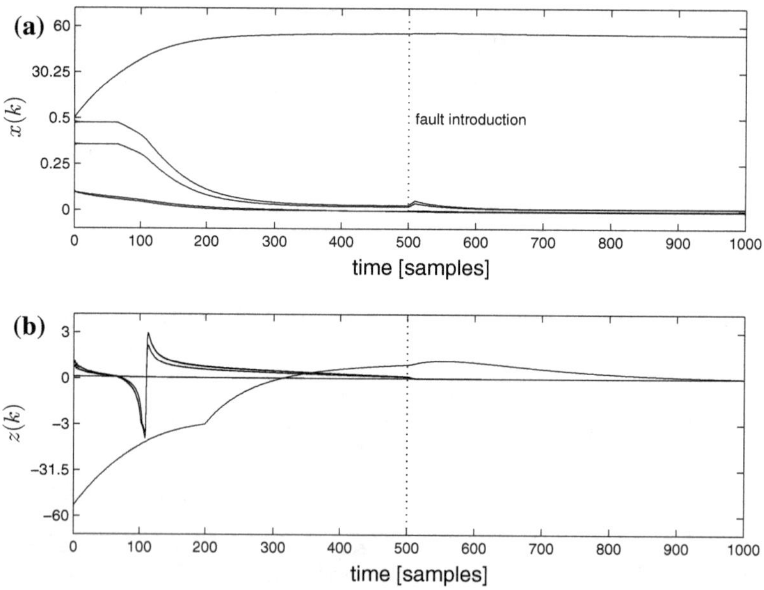

Fig. 5.11 The convergence of the control-system states with a faulty (f_4) system: the original system (**a**), the transformed system (**b**)

Table 5.6 The performance of the LMI solver

Scenario	No fault	f_1	f_2	f_3	f_4
Average time (s)	0.2021	0.2049	0.2059	0.2053	0.2025
Maximum time (s)	0.2760	0.3570	0.3430	0.3050	0.3600
Minimum time (s)	0.1940	0.1940	0.1920	0.1920	0.1890
Average number of iterations	4.0370	4.0350	4.0370	4.0370	4.0370
Maximum number of iterations	8	8	8	8	8
Minimum number of iterations	3	3	3	3	3

The stability of the control system in the case of faults was also verified checking the stability condition (5.44) with the linear matrix inequality method. In order to find the matrix $\boldsymbol{P}$ satisfying the condition (5.44), after each change of the matrix $\boldsymbol{B}$ due to linearization, Yalmip along with SeDuMi is used. The experiment was performed under MATLAB® R2010b (7.11) on a PC with Core 2 Duo T6500 2.1 GHz and 4096 MB RAM (64-bit operating system). The results are presented in Table 5.6. A low iteration number as well as a small average execution time guarantee that such calculations can be accomplished in real time, assuming the sampling time is bigger than 1 s. We note that such an assumption is by all means realistic for the tank unit considered.

5.7 An Experimental Study — A Two-Tank Laboratory System

In the previous example, active FTC based on state-space neural networks was designed for the tank unit simulator implemented in MATLAB®/Simulink® environment [3]. In this study, the developed techniques are used in the context of a laboratory stand, which is a real-life installation [4]. Note that in real-life processes, due to noise or disturbances, fault compensation is much harder to be performed. Yet another difficult problem is related to modelling of the process in the presence of noise and disturbances. Moreover, as the process operates in real-time, the computational burden is also an important problem as there are hard time constraints imposed on the control algorithm. The experiment shows that a proper software implementation of the proposed method is possible, and the approach can be practically used in real industrial process with sampling time equal to 0.5 s. The described methodology uses a robust state-space model of the system and a nonlinear state observer, both designed using artificial neural networks. The designed fault-tolerant control begins to compensate the fault effect by adding to an auxiliary signal to the standard control obtained from the PID controller. This auxiliary control constitutes the additional control loop which can influence the stability of the entire control system. Therefore, stability of the proposed control scheme as based on the Lyapunov direct method is also investigated.

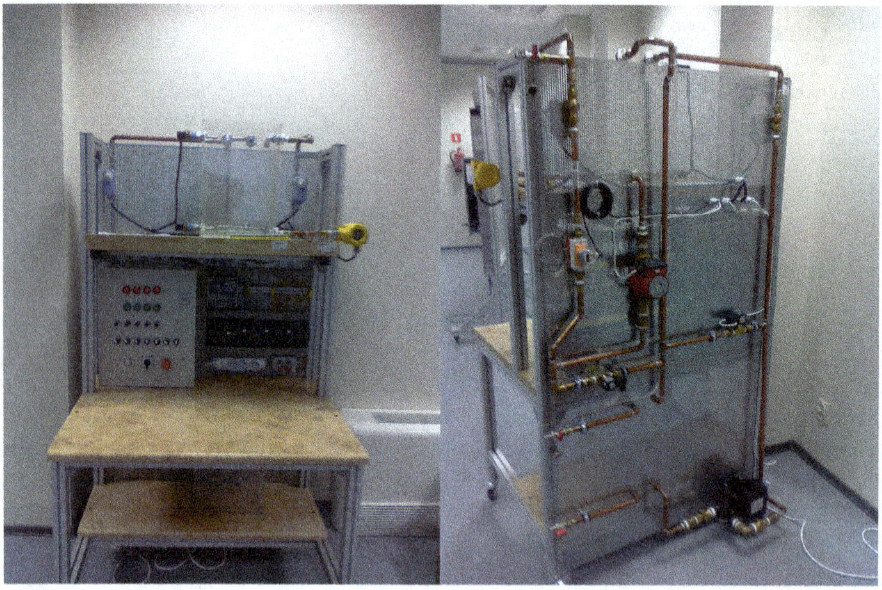

Fig. 5.12 The laboratory installation, front (left) and back (right) sides [4]. ©2014 Elsevier. Reprinted with permission

The object considered in this example is the laboratory installation developed at Research and Development laboratory of the Automation Department in Mazel company, located in Nowa Sól, Poland in the Kostrzyn–Słubice Special Economic Zone. Mazel is a company that is mainly focused on power engineering and automation industry. This work was made possible owing to an agreement between the Faculty of Electrical Engineering, Computer Science and Telecommunications at University of Zielona Góra and Mazel. The entire system consisting of two water tanks and two pumps is presented in Fig. 5.12. The block scheme of the plant is shown in Fig. 5.13. The pumps allow the flow of water between both tanks. The first pump (Pump 1 in Fig. 5.13) ensures water circulation in the upper tank but also speeds up the flow of the water to the lower tank. The second pump (Pump 2 in Fig. 5.13), which is placed at the bottom of the installation, moves water from the lower tank to the upper one and is used to sustain the required water level in the upper tank. In the upper tank, the pressure sensor is installed measuring the water level. There is also a flowmeter measuring the water flow after Pump 1. Various water loops can be implemented by means of a set of different types of valves (solenoid and globe ones). When a heater is installed in the upper tank and a temperature sensor, which measures temperature of circulating water allow the design of more complex control systems. The specification of the process variables is shown in Table 5.7 and the description of valves is presented in Table 5.8. The system can be controlled manually using the control panel, which is placed on the front door of the control cabinet, and in auto mode using the Programmable Logic Controller (PLC). In this case, GE Fanuc VersaMax Micro

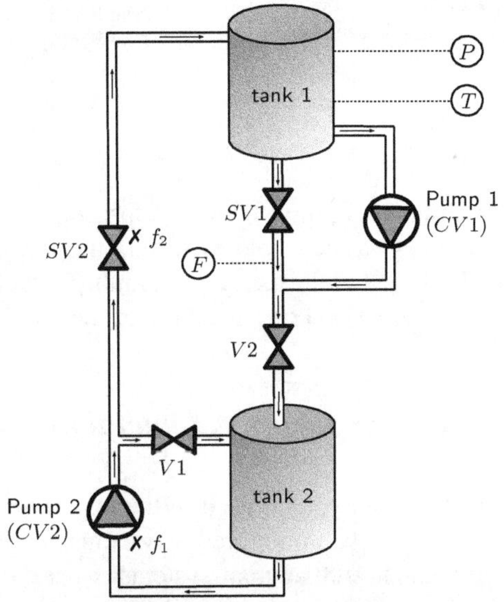

Fig. 5.13 The block scheme of the considered system with fault placement

Table 5.7 The specification of the process variables

Variable	Specification	Range
$CV1$	Pump 1 - control value	0–100%
$CV2$	Pump 2 - control value	0–100%
$CV3$	Heater - control value	0–100%
P	Pressure in tank 1	0–3 kPa
F	Flow in small loop	0.75–9 l/min
T	Water temp. in small loop	0–200 °C

Table 5.8 The specification of the valves

Variable	Specification	Range
$V1$	Valve between tanks	Open–closed
$V2$	Valve between tanks	Open–closed
$SV1$	Globe valve in the small loop	0–100%
$SV2$	Globe valve in the big loop	0–100%

PLC (IC200UDR005) is used in the system. This type of PLC is geared up with three analog signal input-output modules (IC200UEX626) and the ethernet communication module (IC200UEM001). In our experiments, the process was working in the auto mode. All the necessary calculations are made in MATLAB®/Simulink® software and the control signal is processed through OPC Toolbox to the logic controller. PLC generates relevant outputs and sends these to the actuators. The input data from sensor are processed in the same way but in the opposite direction (sensor-PLC-Simulink®). The sampling time with which the data are handled is set to 0.5s. To carry out the experiments with fault detection and accommodation, a number of faulty scenarios were investigated. The considered faults are presented in Table 5.9.

Table 5.9 The specification of the faulty scenarios

Fault	Description	Type
f_1	Pump 1 - power loss	Multiplicative (0.7)
f_2	Blockage in the pipe	Partly closed (70%)
f_3	Fault f_1 and f_2 together	Mixed

To simulate a pump fault (f_1) the control value ($CV1$) was modified. A blockage fault (f_2) was achieved by setting the specific globe valve with the value described in Table 5.9. Note that f_3 is a multiple fault, i.e. where the faults f_1 and f_2 occur simultaneously. The placement of the faults is marked in Fig. 5.13.

5.7.1 The System Assumptions and Configuration

The laboratory stand used in the experiments allows for many different configurations. It can be configured as various simple single-input-single-output systems or multi-input-multi-output systems, where all the actuators and sensors are used. In this study, the structure is selected in the form of multi input single output system where $CV2$ and F are the inputs and P is the output. In such a case, both pumps together with the pressure and flow-meter sensors are used. During the experiments, the valve $V1$ was permanently opened creating continuous connection between the tanks. When Pump 1 is off, water automatically, under the influence of gravity, is transferred from the upper tank to the lower one. When the first pump is running, it realizes two tasks: speeding up the flow of water to the lower tank and also ensuring water circulation in the small loop. This circulation is measured by the flow meter. Pump 2 is controlled by a PI controller and its task is to maintain a constant water level in the upper tank. The controller is designed to maintain a constant amount of water regardless the water amount flowing from the small loop into the upper tank. The upper tank is used to store water and is able to heat water up and receive water from another industrial process. In turn, the lower tank has a dual role: it serves as a water supply and takes in water that was heated or processed in the upper tank.

5.7.2 Modelling

The first step in the fault-tolerant control design is the process modelling. To build a proper model, the training data describing the process under normal operating conditions are required. The input signal should be as informative as possible. The training data were collected in the open-loop control, where the input signal in the form of a random-steps sequence with levels from the interval (0, 100) was used. Each step lasted 25 s for $CV1$ and 100 s for $CV2$. The sampling time was set to 0.5 s.

To form the training set, a sequence of 1500 samples was recorded. The neural network (5.13) was trained for 100 epochs using the Levenberg–Marquardt algorithm and then tested with a different set of data consisting of another 1500 samples.

Table 5.10 The selection of neural network structures [4]. ©2014 Elsevier. Reprinted with permission

Number of neurons	Model order	Training		Testing	
		FPE	SSE	FPE	SSE
3	2	0.05903	171.08	0.0844	244.74
3	4	0.00738	20.659	0.0076	21.380
3	6	0.00148	3.8844	0.0648	169.88
3	8	0.00617	14.794	0.0071	17.139
5	2	0.03160	89.660	0.0840	238.44
5	4	0.00085	2.3640	0.0044	12.345
5	6	0.00115	3.0161	0.0056	14.698
5	8	0.00205	4.9154	0.0033	7.9188
9	2	0.04767	129.58	0.5429	1475.6
9	4	0.00055	1.4283	0.0023	6.0127
9	6	0.00084	2.0617	0.0032	8.0541
9	8	0.00072	1.6966	0.0052	12.322
15	2	0.86188	2196.6	2.5404	6474.8
15	4	0.00060	1.4060	0.0024	5.6726
15	6	0.00119	2.5692	0.0044	9.6348
15	8	0.00093	1.8384	0.0046	9.0884

As described in the Sect. 5.7.1, the model inputs were as follows: the control value $CV2$ and the measured flow F. The model output was the water level in the upper tank as represented by the pressure (P). The neural network (5.13) is defined to be the state observer of the system. However, by setting $\varepsilon(k) = 0$ after the training phase, the observer is converted into the state-space model (5.12). The important problem here is a proper selection of the model structure. The purpose of model selection is to identify a model in order to best fit the data set. On the other hand, the model should be as simple as possible. To evaluate the quality of the modelling, two performance indexes are introduced: a sum of squared errors (SSE), and the Final Prediction Error (FPE) information criterion. The first index is widely used to show how the model fits the data. The second index makes it possible to discard models deemed too complex. Many neural structures with different number of hidden neurons as well as with different order were evaluated. For each evaluated model, the training process was repeated 10 times and the averaged results of the model evaluation are presented in Table 5.10.

The best results are presented surrounded by frames. The best structure was found to be the neural model of the fourth order with nine hidden neurons with the hyperbolic tangent activation function each. It is for this structure that FPE obtains the lowest value for both training and testing sets. As one is able to observe in Table 5.10, a slightly better value of SSE was achieved for the model with fifteen neurons, this model was, however, rejected as too complex.

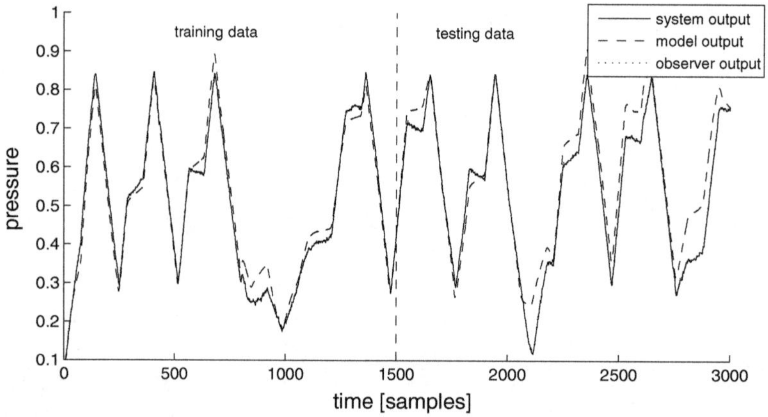

Fig. 5.14 The training results [4]. ©2014 Elsevier. Reprinted with permission

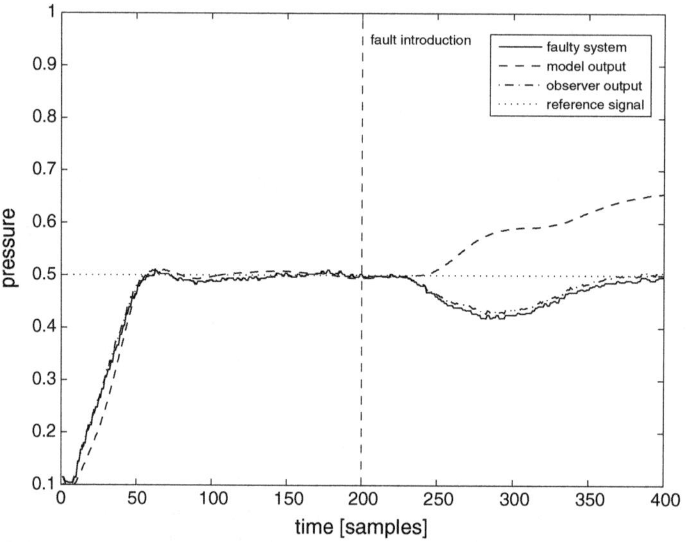

Fig. 5.15 The system operation in the case of the fault f_1 [4]. ©2014 Elsevier. Reprinted with permission

The quality of the modelling for the best neural model is presented in Fig. 5.14. The collected data were divided into two parts for training and testing purposes. In both cases, one sees that the output of the model follows the process output almost immediately, which shows decent generalization capabilities of the model. The output of the state-space observer is also very efficient. The difference between the faulty system and the model behaviour is presented in Figs. 5.15 and 5.16. The model of the system behaves as a healthy system, hence a difference between those outputs, but the observer of the system follows the system even in the case of a fault.

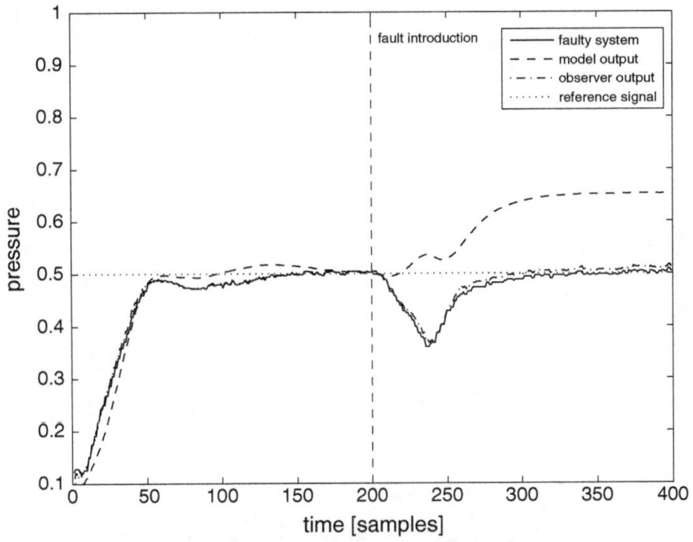

Fig. 5.16 The system operation in the case of the fault f_2 [4]. ©2014 Elsevier. Reprinted with permission

5.7.3 Fault Detection

The purpose of the fault detection block designed as described in Sect. 5.4 is to monitor the working condition of the plant and trigger the compensation when a fault is detected. The objective of the control system working with the PI controller was to keep the constant water level in the upper tank represented by the pressure equal to 1.5 kPa, which is half of the maximum water level in the upper tank. The desired water level in the upper tank is achieved through controlling only the $CV2$ value. The second input was not controllable (due to minor impact on the water level) and was rigidly set to the value of 3 l/min by means of a separate PID controller. Each fault listed in Table 5.9 was simulated at the $200th$ time instant which was the 100th second of the simulation. In Figs. 5.17, 5.18 and 5.19 the output of system and uncertainty bounds without fault compensation are presented. When the trajectory of the system output crosses either the lower or the higher uncertainty bounds, the fault alarm is turned on. The detection time for the faulty scenarios f_1, f_2 and f_3 is equal to 212, 248 and 216 time instants, respectively.

5.7.4 Fault Compensation

In order to check the proposed fault accommodation approach, several experiments were carried out investigating the behaviour of the fault-tolerant control system by

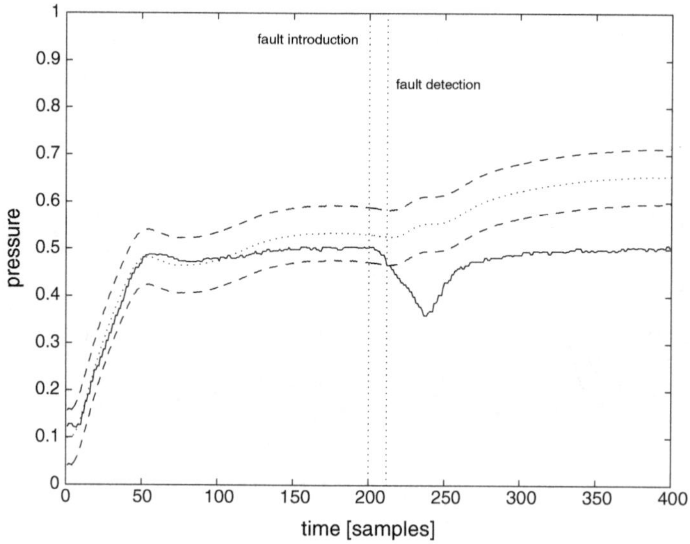

Fig. 5.17 Model error modelling; the system output (solid) and uncertainty bands (dashed) in the case of a faulty scenario f_1 [4]. ©2014 Elsevier. Reprinted with permission

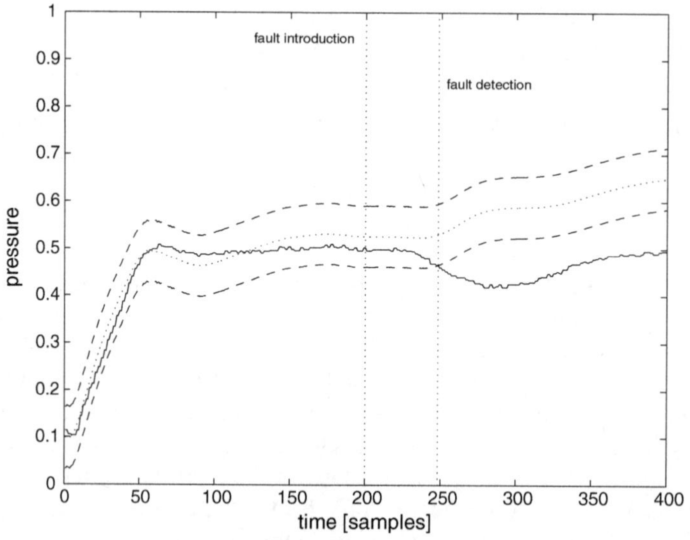

Fig. 5.18 Model error modelling; system output (solid) and uncertainty bands (dashed) in the case of a faulty scenario f_2 [4]. ©2014 Elsevier. Reprinted with permission

introducing several faults in the system. When the detection block (described in Sect. 5.4) triggered the fault alarm signal, the system changed its state from the nominal to the faulty one. In the nominal conditions, the value of the additional

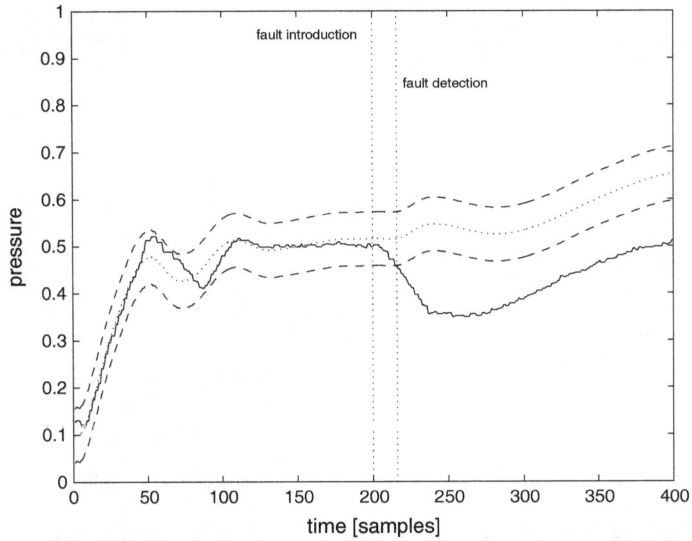

Fig. 5.19 Model error modelling; system output (solid) and uncertainty bands (dashed) in the case of a faulty scenario f_3 [4]. ©2014 Elsevier. Reprinted with permission

control $\boldsymbol{u}^{\mathrm{fa}}$ is set to 0, however in the case of an occurrence of a fault, its value is changed according to (5.10) and it can be then applied to in the context of control if the stability condition (5.50) is satisfied. The achieved results are also compared with a different fault-tolerant approach of the passive type [19, 20]. In this alternative fault-tolerant scheme, a reconfiguration of the controller is simply done by switching between the already designed PID controllers. Each of the controllers is designed for one of the predefined faulty situation. Such an approach demonstrates a very good control efficiency in cases when known (anticipated) faults affect the system.

Figures 5.20, 5.21 and 5.22 present the behaviour of the fault-tolerant system in the case of the faults listed in Table 5.9. Each chart presents the output of the healthy system (the black solid line), the output of the faulty system without compensation (with PID only) (the blue solid line), the output of the passive FTC (the green solid line) and the output of the compensated system (the proposed solution) (the red solid line). In turn, fault-tolerant control results are given in Table 5.11. The presented results are calculated for 200 samples, starting from the time instant when the fault is introduced, until the time instant when the simulation is finished. The simulation stops when the fault is compensated and the system follows the reference trajectory again. The improvement for the each faulty scenario is calculated as follows:

$$\text{Improvement (\%)} = \left(1 - \frac{\text{SSE}^{ftc}}{\text{SSE}^{noftc}}\right) 100\%. \tag{5.51}$$

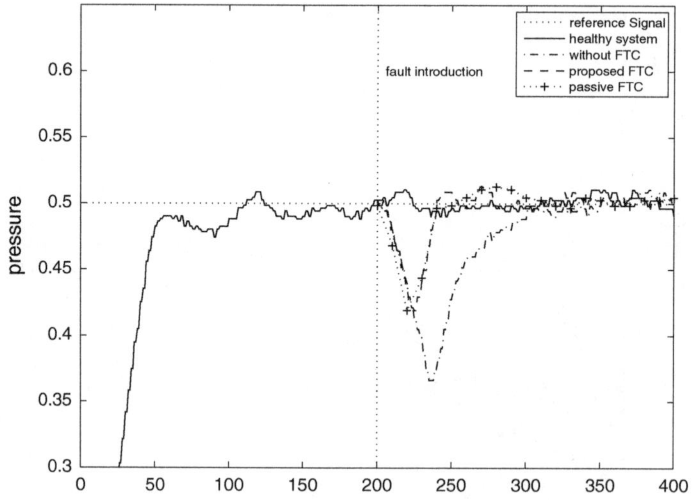

Fig. 5.20 Fault accommodation for the faulty scenario f_1 [4]. ©2014 Elsevier. Reprinted with permission

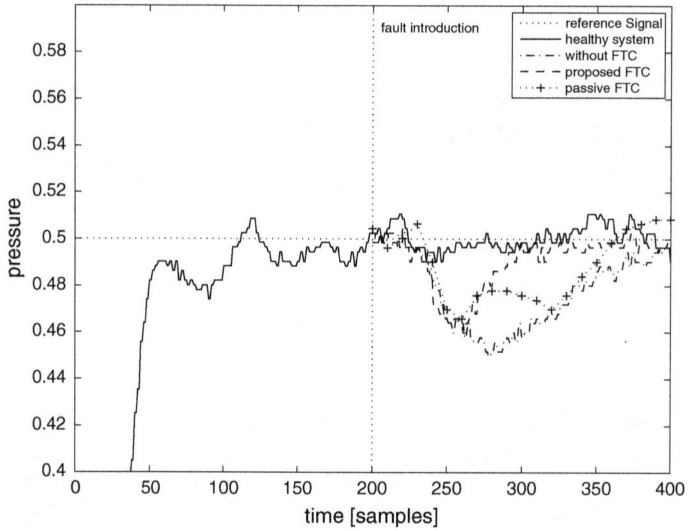

Fig. 5.21 Fault accommodation for the faulty scenario f_2 [4]. ©2014 Elsevier. Reprinted with permission

The achieved results clearly show that by using the proposed active FTC scheme, every fault can be compensated satisfactorily, in contrast to the situation of the standard control system without fault-tolerance properties. The achieved improvement

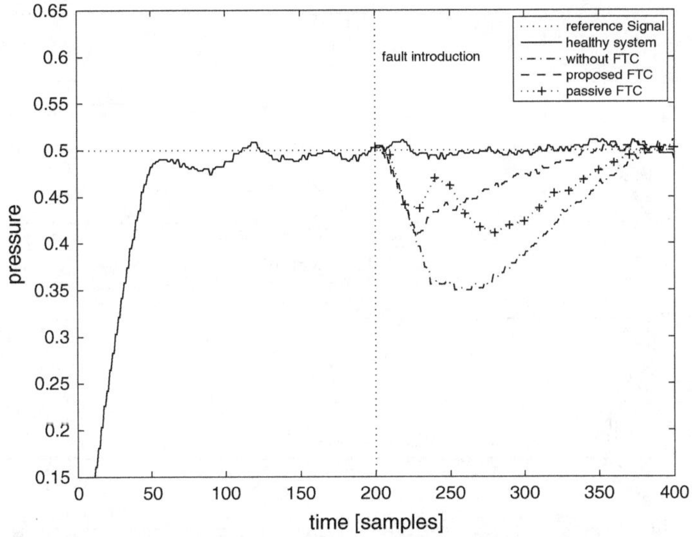

Fig. 5.22 Fault accommodation for the faulty scenario f_3 [4]. ©2014 Elsevier. Reprinted with permission

Table 5.11 The results of the experiments in form of SSE and percentage index [4]. ©2014 Elsevier. Reprinted with permission

Scenario	f_1	f_2	f_3
SSE without FTC	0.4126	0.1476	1.7155
SSE with passive FTC	0.1095	0.0681	0.5153
Improvement (%)	73.4631	53.8631	69.9602
SSE with proposed FTC	0.0898	0.0430	0.2952
Improvement (%)	78.2449	70.8857	82.7918

varies from 70.8857% (the fault f_2) to 82.7918 (the fault f_3). The results are as expected owing to the fact that any fault-tolerant approach should work better that a control system without inherent fault-tolerant capabilities. Therefore, in order to truly evaluate the effectiveness of the proposed approach, it was compared to the alternative passive FTC approach (the results presented in the third and the fourth row of Table 5.11). The achieved improvement, in contrast to the standard control systems, varies from 53.8631% (the fault f_2) to 73.4631 (the fault f_1). Comparing both considered FTC approaches, one can state that the proposed active FTC compensates all of the faults in a better way. The relative improvement is: 4.78% for the fault f_1, 17.02% for the fault f_2 and 12.83% for the fault f_3.

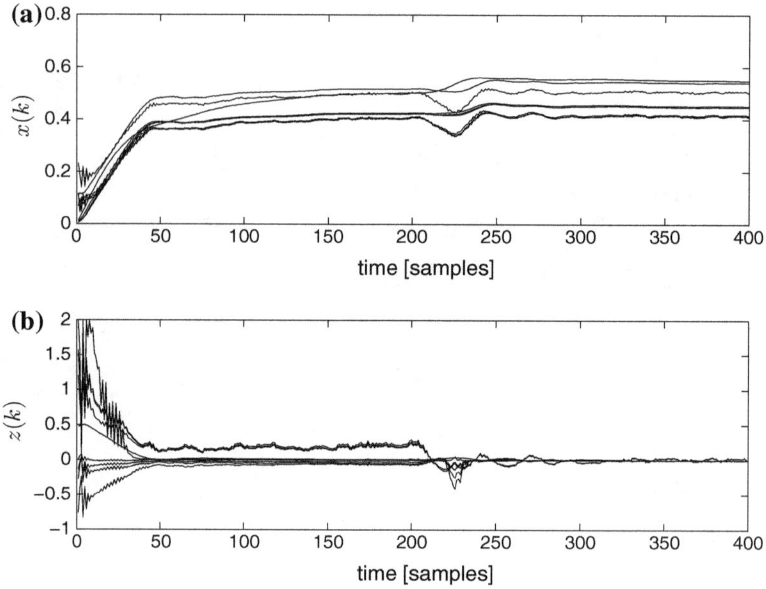

Fig. 5.23 The convergence of the control system states with faulty (f_1) system: the original states (**a**), the transformed states (**b**) [4]. ©2014 Elsevier. Reprinted with permission

5.7.5 *Stability Analysis*

In case of the fault-tolerant control scheme examined in this chapter, the correction of the control signal constitutes the additional control loop in the operation of the system. This can disturb the stability of the control system. As shown in Sect. 5.5, this can be very efficiently checked with the Lyapunov direct method. We also take into account the convergence of the system states of the original system (5.37). The results achieved in the case of the faults f_1, f_2 and f_3 are shown in Fig. 5.23a, 5.24a and 5.25a, respectively. In Figs. 5.23b, 5.24b and 5.25b, the states of the transformed autonomous system (5.43) are presented. All the transformed states converge to zero, which means that the control system is stable. The stability of the control system in the case of faults was also verified checking the stability condition (5.50) using linear matrix inequalities (LMIs). In order to find the matrix $\boldsymbol{P}$ satisfying the condition (5.44), after each change of the matrix $\boldsymbol{B}$ due to linearization, the Yalmip together with SeDuMi is used. The experiment was performed using MATLAB® R2012a (7.14) on a PC with Core i7-2670QM 2.2-3.1 GHz and 8192 MB RAM (64-bit operating system). The results are presented in Table 5.12. In Figs. 5.26a, 5.27a and 5.28a we present the iteration number required to find a solution to calculate the

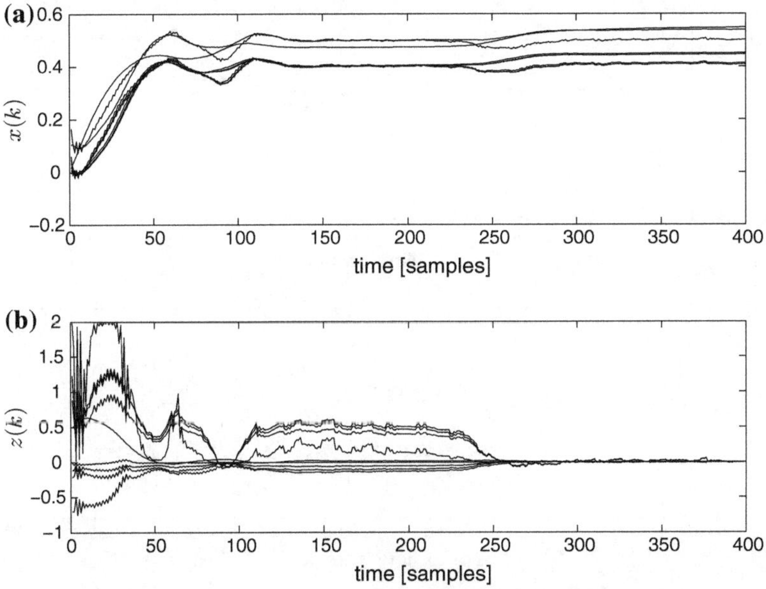

Fig. 5.24 The convergence of the control system states with faulty (f_2) system: the original states (**a**), the transformed states (**b**) [4]. ©2014 Elsevier. Reprinted with permission

Table 5.12 The performance of the LMI solver [4]. ©2014 Elsevier. Reprinted with permission

Faulty scenario	f_1	f_2	f_3
Avg. time (s)	0.1577	0.1684	0.1727
Max time (s)	0.2920	0.3500	0.3910
Min time (s)	0.1150	0.1120	0.1160
Avg. num of iter.	9.3075	9.3775	9.6525
Max num of iter.	11	12	13
Min num of iter.	8	8	9

matrix $\boldsymbol{P}$ at each time instant of the simulation, while Figs. 5.26b, 5.27b and 5.28b show the computation time for each sample. The low iteration number as well as small average execution time guarantee that such calculations can be accomplished in real time, assuming the sampling time bigger than 0.5 s. Such an assumption is by all means justified in the case of the considered system.

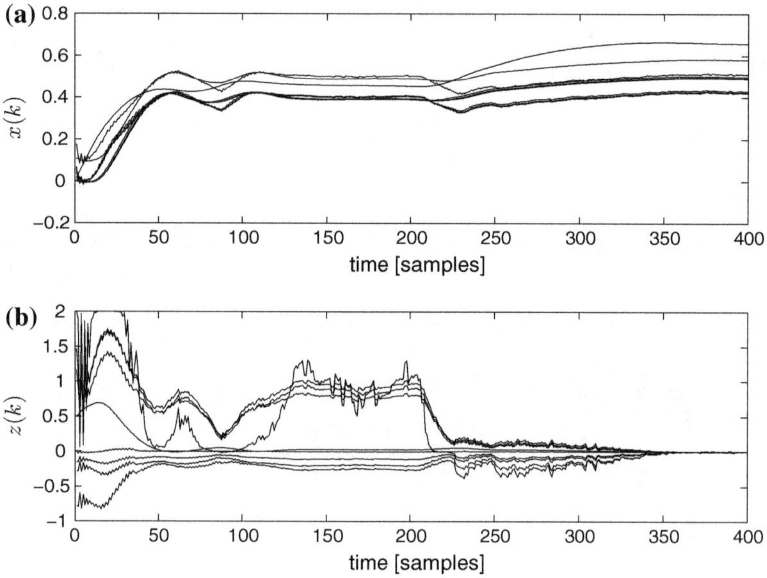

Fig. 5.25 The convergence of the control system states with faulty (f_3) system: the original states (**a**), the transformed states (**b**) [4]. ©2014 Elsevier. Reprinted with permission

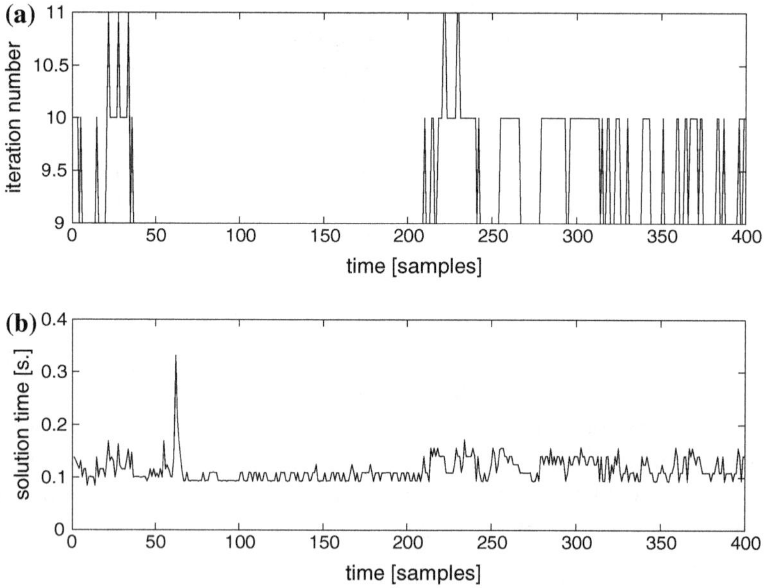

Fig. 5.26 The results of solving LMI in the f_1 faulty scenario: the number of iteration (**a**), solving time (**b**)

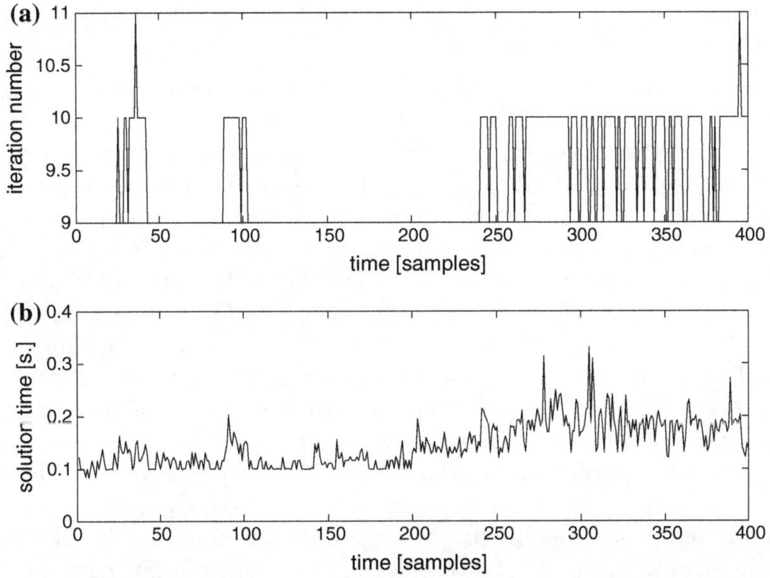

Fig. 5.27 The results of solving LMI in the f_2 faulty scenario: the number of iteration (**a**), solving time (**b**)

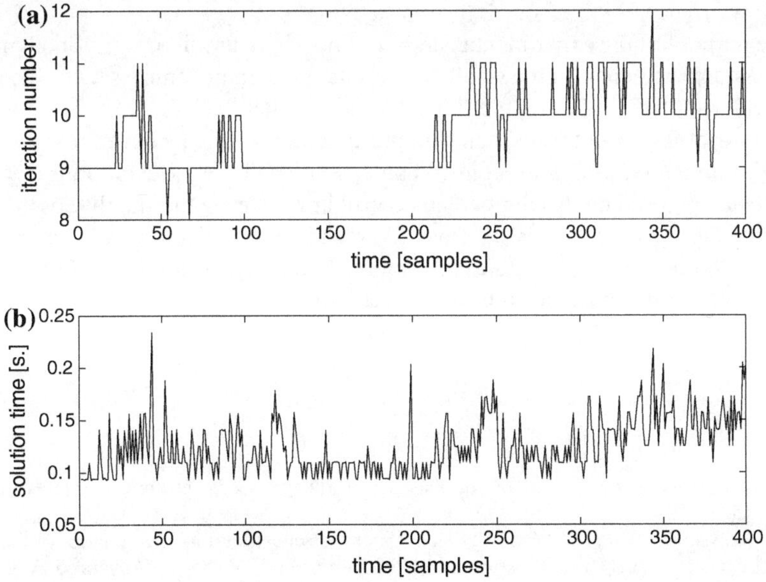

Fig. 5.28 The results of solving LMI in the f_3 faulty scenario: the number of iteration (**a**), solving time (**b**)

5.8 Conclusions

The purpose of this chapter was to design and examine fault-detection and accommodation system based on a robust state-space neural model for a fluid-flow and pressure-control laboratory stand. Fault compensation was carried out applying the modified control law derived using the instantaneous linearization of the already-trained nonlinear state-space model of the system. The proposed method makes it possible to improve the work of the considered nonlinear system in the case of faults. The obtained experimental results confirm that the model in the form of the state-space neural network can be effectively and easily used to minimize the residuum defined as the difference between the outputs of the nominal and a faulty system. The applied linearization technique is very simple and not time consuming. As was shown in Table 5.11, the faults were compensated about $70 - 82\%$ better in comparison to the classical PI controller. Moreover, the proposed active FTC works better than an alternative passive one based on controller switching. For the considered set of faults, the relative improvement ranges from 4.78% to even 12.08%.

The efficiency of the fault compensation depends on the quality of the fault-detection module. Using the robust model of the system in this task allows one to detect faults quickly and reliably, it also means that the procedure of the fault accommodation can be used quicker. The results are therefore quite satisfactory, yet we believe that they can still be improved by developing a better model of the system (which will be a subject of our future work).

The purpose of the experimental study was also to determine, verify (both numerically and experimentally) the stability of the fault-tolerant control scheme designed for a real installation. The reported results showed that the Lyapunov direct method is a very useful tool in this aspect. The presented experiments have proved that the proposed active fault-tolerant control can ensure stable work conditions. Defining constraints imposed on the augmented control law assuring the stability of the fault-tolerant control system in any case would be an interesting problem to be explored. Such constraints can be defined extracting the knowledge from the stability criteria. This problem constitutes our future research directions in this area.

References

1. Chen, J., Patton, R.J.: Robust Model-based Fault Diagnosis for Dynamic Systems. Kluwer, Berlin (1999)
2. Czajkowski, A., Patan, K.: Stability analysis of the robust fault tolerant control of the boiler unit using state space neural networks. In: 9th European Workshop on Advanced Control and Diagnosis, Budapest, Hungary, p. 8 (2011)
3. Czajkowski, A., Patan, K., Korbicz, J.: Stability analysis of the neural network based fault tolerant control for the boiler unit. In: Rutkowski, L., Korytkowski, M., Scherer, R., Tadeusiewicz, R., Zadeh, L., Zurada, J. (eds.) Artificial Intelligence and Soft Computing: 11th International Conference, ICAISC 2012, Zakopane, Poland, 29 April–3 May 2012, Proceedings, Part II, vol. 7268, pp. 548–556. Springer, Berlin (2012)

4. Czajkowski, A., Patan, K., Szymański, M.: Application of the state space neural network to the fault tolerant control system of the PLC-controlled laboratory stand. Eng. Appl. Artif. Intell. **30**, 168–178 (2014)
5. Korbicz, J., Kościelny, J., Kowalczuk, Z., Cholewa, W. (eds.): Fault Diagnosis. Models, Artificial Intelligence, Applications. Springer, Berlin (2004)
6. Kou, P., Zhou, J., Wang, C., Xiao, H., Zhang, H., Li, C.: Parameters identification of nonlinear state space model of synchronous generator. Eng. Appl. Artif. Intell. **24**(7), 1227–1237 (2011)
7. Mendonça, L., Sousa, J., Sá da Costa, J.: Fault tolerant control using a fuzzy predictive approach. Expert Syst. Appl. **39**(12), 10630–10638 (2012)
8. Nørgaard, M., Ravn, O., Poulsen, N., Hansen, L.: Networks for Modelling and Control of Dynamic Systems. Springer, London (2000)
9. Noura, K., Theilliol, D., Ponsart, J.C., Chamseddine, A.: Fault Tolerant Control Systems. Design and Practical Applications. Advanced in Industrial Control. Springer, London (2009)
10. Patan, K.: Artificial Neural Networks for the Modelling and Fault Diagnosis of Technical Processes. Lecture Notes in Control and Information Sciences. Springer, Berlin (2008)
11. Patan, K., Parisini, T.: Identification of neural dynamic models for fault detection and isolation: the case of a real sugar evaporation process. J. Process Control **15**, 67–79 (2005)
12. Patan, K., Witczak, M., Korbicz, J.: Towards robustness in neural network based fault diagnosis. Int. J. Appl. Math. Comput. Sci. **18**(4), 443–454 (2008)
13. Polycarpou, M., Vemuri, A.T.: Learning methodology for failure detection and accommodation. IEEE Control Syst. Mag. **15**, 16–24 (1995)
14. Qian, M., Jiang, B., Xu, D.: Fault tolerant tracking control scheme for UAV using dynamic surface control technique. Circuits Syst. Signal Process. **31**(5), 1713–1729 (2012)
15. Reinelt, W., Garulli, A., Ljung, L.: Comparing different approaches to model error modeling in robust identification. Automatica **38**, 787–803 (2002)
16. Schröder, D.: Intelligent Observer and Control Design for Nonlinear Systems. Springer Science and Business Media, Berlin (2000)
17. Tornil-Sin, S., Ocampo-Martinez, C., Puig, V., Escobet, T.: Robust fault detection of nonlinear systems using set-membership state estimation based on constraint satisfaction. Eng. Appl. Artif. Intell. **25**(1), 1–10 (2012)
18. Verron, S., Tiplica, T., Kobi, A.: Fault diagnosis of industrial systems by conditional gaussian network including a distance rejection criterion. Eng. Appl. Artif. Intell. **23**(7), 1229–1235 (2010)
19. Yang, H., Jiang, B., Staroswiecki, M.: Supervisory fault tolerant control for a class of uncertain nonlinear systems. Automatica **45**(10), 2319–2324 (2009)
20. Yang, H., Jiang, B., Cocquempot, V., Lu, L.: Supervisory fault tolerant control with integrated fault detection and isolation: a switched system approach. Appl. Math. Comput. Sci. **22**(1), 87–97 (2012)
21. Zhang, Q.: Using wavelet networks in nonparametric estimation. IEEE Trans. Neural Netw. **3**(2), 227–236 (1997)
22. Zhang, Y.: Active fault-tolerant control systems: integration of fault diagnosis and reconfigurable control. In: Korbicz, J., Patan, K., Kowal, M. (eds.) Fault Diagnosis and Fault Tolerant Control. Challenging Problems of Science - Theory and Applications: Automatic Control and Robotics, pp. 21–41. Academic Publishing House EXIT, Warsaw (2007). ISBN: 978-83-60434-32-1

Chapter 6
Iterative Learning Control

6.1 Introduction

The fact that modern engineering systems are increasingly more complex means that control system design has to meet ever growing quality demands. What is required is more accurate control strategies able to adapt to changes in uncertain environmental conditions. Recall that many factors can contribute to uncertainty in system modelling [22], e.g. measurement noise, nonlinear plant dynamics, modelling errors, etc. The classical approaches to control system design dedicated to linear systems are either incapable of meeting the required quality levels or are not easily implementable in cases of more complex system dynamics. Therefore, new modern robust control methods have emerged and have become the subject of intensive research. In the context of repetitive processes, it is the iterative learning control technique that became especially attractive [2]. Such processes can be frequently encountered in industrial production situations. Their main goal is to replicate certain operations in consecutive trials in order to ensure that the system response is accurately following some arbitrarily given reference trajectory. Often, the typical industrial controller replicates errors at consecutive process trials without any improvement [18] thus outputting the same behavior of the tracking error in each trial (e.g. the overshoot and oscillations).

ILC as an effective data-driven method designed to aid in an improvement of a control signal, originally proposed in late 1970s. It has nowadays grown to a separate field within control theory [1, 27]. This machine learning technique proved its usefulness in many real-world engineering applications, including industrial robots [2, 18, 23, 26], batch processes [13, 14], computer numerical control machine tools [11], injection-molding machines [8, 9], rapid thermal processing [30], combustion processes [12] and more recently, health care systems [6, 7].

On the other hand, the practitioners are especially interested in control systems having a simple structure, in order to fulfill the requirements imposed on the cost and maintenance process control. The iterative learning control (ILC) approach is

© Springer Nature Switzerland AG 2019

K. Patan, *Robust and Fault-Tolerant Control*, Studies in Systems, Decision and Control 197, https://doi.org/10.1007/978-3-030-11869-3_6

in some respect effective data-driven method that is based on on-line or off-line input/output data.

Although ILC is a robust control technique with numerous engineering applications, it is still not able to compensate random disturbances. This leads to difficulties in learning performance and thus potentially to a slow convergence of the tracking error. In addition to this, a great majority of the existing ILC schemes still require a mathematical model of a plant (or its nonparametric equivalent) in order to ensure a proper design of the control signal. The problem becomes even more difficult when the model cannot be easily inverted, which leads to highly non-trivial control design situations. In this context, artificial neural networks have been intensively studied during the last two decades and proved their usefulness in control theory, especially in the area of system modelling and plant control [19]. Neural networks provide an appealing and flexible alternative to the classical methods, as they can handle nonlinear problems [10, 16]. They also play a crucial role in cases when it is difficult (or even impossible) to derive a mathematical model of the plant. Yet another attractive property of neural networks is their self-learning capability. A neural network can extract the system features from historical training data using a learning algorithm, requiring little or no a priori knowledge about the process. These properties make the ANNs especially suited for being incorporated in the general ILC scheme.

Therefore, the main idea here is to adopt ANNs to accurately identify the mathematical model of a plant, based on the available measurement data and to properly train the neural controller. In this way, it is possible to indirectly increase the quality of control design and increase the error convergence rate via model tuning. This makes it possible to address two important issues. The first is related to building the accurate model of the nonlinear plant based on the measurements from the previous trials in order to reduce the modelling uncertainty, hence improving the quality of control. The second issue is related to the fact that an inverted neural model can be obtained with relative ease, making it possible to effectively design a feed-forward ILC controller.

Obviously, applying ANNs for the purposes of ILC is not a new idea, however the main contribution of this work is to propose a special ILC scheme with an additional neural network for gradient estimation in the context of training of the neural controller. Also, we have derived and discussed the necessary conditions for achieving neural controller stability. The theoretical part is made more accessible by showing its practical application in two example problems relating to a repetitive control of a pneumatic servomechanism and a magnetic suspension system.

Notation

We use $\mathbb{N}$ and $\mathbb{R}$ to denote the sets of non-negative integers and real numbers, respectively. The n-dimensional Euclidean vector space is denoted by $\mathbb{R}^n$, and the Euclidean matrix space of real matrices with n rows and k columns is denoted by $\mathbb{R}^{n\times k}$. We will write $\mathbb{N}_{<\ell}$ as set of all non-negative integers less than ℓ, i.e. $\mathbb{N}_{<\ell} = \{0, 1, \ldots, \ell - 1\}$. For $\boldsymbol{A} \in \mathbb{R}^{\ell\times m}$ we define the svec operator as $\mathrm{svec}(\boldsymbol{A}) = (a_{11}, a_{21} \ldots, a_{\ell 1}, \ldots, a_{1m}, \ldots, a_{\ell m})^{\mathsf{T}}$, which is the column vector of size ℓm obtained by stacking the columns of the $\boldsymbol{A}$. The symbols $\mathbf{1}$ and $\mathbf{0}$ denote vectors whose all components are one and zero, respectively.

6.2 Iterative Learning Control Design

Let us consider the discrete-time SISO system:

$$y_p(k) = P(z)u_p(k) + d(k), \tag{6.1}$$

where k is the discrete-time index, p stands for the trial index, z is the forward time shift operator, $P(z)$ is a proper rational function of z of relative degree of m, $u_p(k)$, $k = 0, \ldots, N-1$ is the input, $y_p(k)$ $k = m, \ldots, N-m+1$ is the plant output, and $d(k)$, $k = m, \ldots, N-m+1$ is the exogenous input which repeats each trial. It is assumed that $P(z)$ is asymptotically stable. If $P(z)$ is not stable it should be stabilized with a feedback controller. The tracking error is defined as $e_p(k) = y_r(k) - y_p(k)$, with $y_r(k)$, $k = m, \ldots, N-m+1$ stands for the reference.

A common form of the ILC algorithm is:

$$u_{p+1}(k) = Q(z)\left[u_p(k) + L(z)e_p(k+1)\right], \tag{6.2}$$

where $Q(z)$ and $L(z)$ are the so-called Q-filter and learning function, respectively. The design process of ILC scheme boils down to proper selection of $Q(z)$ and $L(z)$ in order to satisfy the convergence and stability properties of a iterative learning control scheme. For the commonly used design procedures in terms of linear systems the interested reader is referred to the survey paper [2].

In the case of unstable systems, as well as for the purposes of accommodating of the non-repeated disturbances, it is advisable to combine learning and feedback controllers. Such a fusion leads to the so-called current-iteration ILC, represented as:

$$u_{p+1}(k) = Q(z)\left[u_p(k) + L(z)e_p(k+1)\right] + C(z)e_{p+1}(k), \tag{6.3}$$

where $C(z)$ represents the dynamics of the feedback controller. The control law (6.3) refers to the parallel architecture of current-iteration ILC is illustrated in Fig. 6.1a. A slight modification of this scheme can be achieved by feeding the learning controller with its own output from the previous trial, instead of augmented control, as follows:

$$u_{p+1}(k) = w_{p+1}(k)C(z)e_{p+1}(k), \tag{6.4}$$

where

$$w_p(k) = Q(z)\left[u_p(k) + L(z)e_p(k+1)\right]. \tag{6.5}$$

This kind of ILC scheme is shown in Fig. 6.1b. ILC can also be arranged in a serial manner, as depicted in Fig. 6.2. In this case, the signal provided by the learning controller is as follows:

$$u_{p+1}(k) = Q(z)\left[u_p(k) + L(z)e_p(k+1)\right]. \tag{6.6}$$

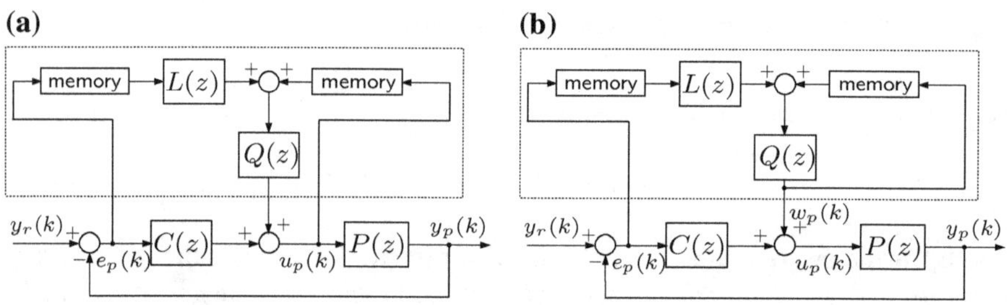

Fig. 6.1 Parallel architectures of current-iteration ILC

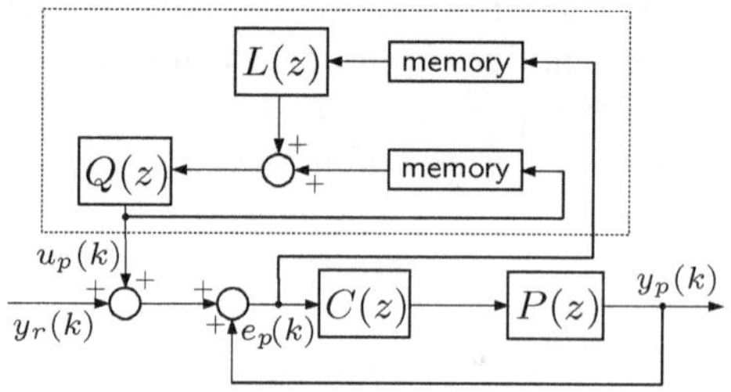

Fig. 6.2 Serial architectures of current-iteration ILC

The name of the scheme comes from the fact that the ILC control is applied to the reference before the feedback loop. It should be pointed out that setting $u_p(k)$ to zero in both cases results in the classical feedback control. The majority of existing ILC approaches are devoted to linear systems [2, 17, 24] and plenty of designing methods have been described to date [2]:

- PD-type and tunable realizations,
- plant inversion methods,
- H_∞ methods,
- quadratically optimal design.

However, industrial plants are mostly characterized by a nonlinear behaviour. Extensions to nonlinear systems have also been proposed [3, 29] nonetheless, these usually use a linear form of the learning controller [25, 28].

The real usefulness of ILC is observed when it is required to control the transient response behaviour of nonlinear or time-varying systems. In such a case, it makes sense to consider the learning controllers that have the ability to adapt to such a system structure. Both objectives can be achieved through the application of artificial neural networks. Neural networks can easily deal with nonlinear problems as well as adapt to changing working conditions of a plant. The first proposals to employ neural networks in the context of ILC originate from early 1990s [15]. Surprisingly, in spite of attractiveness of this idea, one notices this research field remained underdeveloped. In the paper [5], the authors proposed ILC based on two recurrent neural networks representing a plant and the controller, respectively. However, the authors focused on network training without providing any results in terms of a convergence analysis.

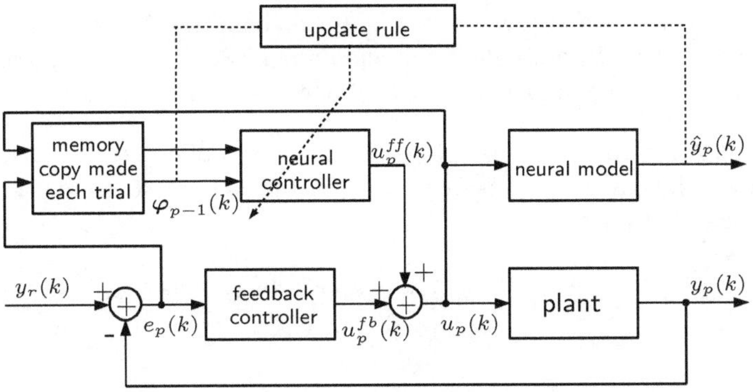

Fig. 6.3 A general structure of iterative learning control based on neural networks

In the work [4], two neural networks were applied to represent nonlinearities of the plant with the control law derived in terms of feedback linearization. However, the applicability of the proposed approach is limited to a specific class of nonlinear systems. In [28], neural networks are used only for modelling of a plant while the learning controller has a classic linear form.

The main idea considered here is to apply a neural network to provide the realization of the function f, which, in fact, is implicitly an inverted model of the plant. This can be done using the historical data recorded in the plant. However, the training process of this neural controller requires an accurate estimation of the system response sensitivities with respect to the neural network parameters. Taking this into account, in general, an analytical description of the plant is unknown, especially in the case of nonlinear plants. An estimation of the system response sensitivities can be achieved by means of another neural network that models the controlled plant. A general ILC scheme is depicted in Fig. 6.3, which is based on the existing feedback controller and it is called current-iteration ILC. The control law has the form:

$$u_p(k) = u_p^{fb}(k) + u_p^{ff}(k), \tag{6.7}$$

where $u_p^{fb}(k)$ is the control law provided by the feedback controller and $u_p^{ff}(k)$ is the learning component provided in the feed-forward manner.

6.3 Static Learning Controller

Let us consider a class of scalar nonlinear discrete-time systems described by:

$$\begin{aligned} \boldsymbol{x}_p(k+1) &= g(\boldsymbol{x}_p(k), \boldsymbol{u}_p(k)), \quad k = 0, \ldots, N-1 \\ \boldsymbol{y}_p(k) &= \boldsymbol{C}\boldsymbol{x}_p(k) \end{aligned}, \tag{6.8}$$

where the integer $p \in \mathbb{N}$ stands for the trial (cycle number), N is a length of a trial, $\boldsymbol{x}_p(k)$, $\boldsymbol{u}_p(k)$ and $\boldsymbol{y}_p(k)$ are state, input and output of a system along the pth trial and g is a nonlinear function. Let us make the following assumptions:

Assumption A1 Let $\boldsymbol{y}_r(k)$ be a reference trajectory defined over a discrete time $k \in \mathbb{N}_{<N}$, which is assumed to be realizable, that is there exists a unique $\boldsymbol{u}_r(k)$ and an initial state $\boldsymbol{x}_r(0)$, i.e.

$$\begin{aligned} \boldsymbol{x}_r(k+1) &= g(\boldsymbol{x}_r(k), \boldsymbol{u}_r(k)) \\ \boldsymbol{y}_r(k) &= \boldsymbol{C}\boldsymbol{x}_r(k) \end{aligned} . \tag{6.9}$$

Assumption A2 The same initial condition holds for all trials, i.e.

$$\forall p \;\; \boldsymbol{x}_p(0) = \boldsymbol{x}_r(0). \tag{6.10}$$

Assumption A3 The nonlinear function g satisfies the global Lipschitz condition:

$$\|g(\boldsymbol{x}_1, \boldsymbol{u}_1) - g(\boldsymbol{x}_2, \boldsymbol{u}_2)\| \leq L\left(\|\boldsymbol{x}_1 - \boldsymbol{x}_2\| + \|\boldsymbol{u}_1 - \boldsymbol{u}_2\|\right), \tag{6.11}$$

where $L > 0$ stands for the Lipschitz constant.

We consider a typical scheme, where the data from an entire previous trial is used in order to derive a new control input [1, 27]. In order to simplify the presentation, we focus on the single-input single-output systems, i.e. $n_u = n_y = 1$. This simplification, however, does not alter the generality of the considerations as all the following results can be generalized to the multi-input multi-output case in a rather straightforward manner. Thus, the ultimate goal of an iterative control algorithm represented by some function f is to follow the reference trajectory to a very high accuracy, i.e. to converge uniformly with the tracking error $e_p(k) = y_r(k) - y_p(k)$ as close to zero as possible.

6.3.1 A Model of the System

Let us consider a state-space neural network model of the nonlinear system (6.8):

$$\begin{aligned} \hat{\boldsymbol{x}}_p(k+1) &= \hat{g}(\hat{\boldsymbol{x}}_p(k), \boldsymbol{u}_p(k)) \\ \hat{\boldsymbol{y}}_p(k) &= \boldsymbol{C}\hat{\boldsymbol{x}}_p(k) \end{aligned} , \tag{6.12}$$

where $\hat{\boldsymbol{x}}_p \in \mathbb{R}^{n_x}$ is the model state, $\boldsymbol{u}_p \in \mathbb{R}^{n_u}$ and $\hat{\boldsymbol{y}}_p \in \mathbb{R}^{n_y}$ are the input and output, respectively, and the nonlinear function $\hat{g}(\cdot, \cdot)$ has the form:

$$\hat{g}(\hat{\boldsymbol{x}}_p(k), \boldsymbol{u}_p(k)) = \boldsymbol{V}_2\sigma(\boldsymbol{V}_1^x\hat{\boldsymbol{x}}_p(k) + \boldsymbol{V}_1^u\boldsymbol{u}_p(k) + \boldsymbol{\beta}_1) + \boldsymbol{\beta}_2), \tag{6.13}$$

where $\boldsymbol{V}_1^u \in \mathbb{R}^{v_m \times n_u}$, $\boldsymbol{V}_1^x \in \mathbb{R}^{v_m \times n_x}$ and $\boldsymbol{V}_2 \in \mathbb{R}^{n_x \times v_m}$ are input-to-hidden, state-to-hidden and hidden-to-output layers weight matrices, respectively, $\boldsymbol{\beta}_1 \in \mathbb{R}^{v_m}$ and $\boldsymbol{\beta}_2 \in \mathbb{R}^{n_x}$ are bias vectors, $\sigma : \mathbb{R}^{v_m} \to \mathbb{R}^{v_m}$ is the activation function of the hidden layer and v_m stands for the number of hidden neurons. Finally $\boldsymbol{C} \in \mathbb{R}^{n_y \times n_x}$ stands for the output (observation) matrix. The structure of the state-space neural network is shown in Fig. 2.19. The network consists of two hidden layers of neurons. The first layer is composed of the neurons with nonlinear activation functions σ, while the second layer includes linear units. The common choice of the nonlinear activation function is to use the hyperbolic tangent function $\sigma(x) = \tanh(x)$ satisfying the following conditions:

$$\begin{aligned}
&\text{(i)} && \sigma(x) \to \pm 1 \text{ as } x \to \pm\infty,\\
&\text{(ii)} && \sigma(x) = 0 \text{ at a unique point } x = 0,\\
&\text{(iii)} && \sigma'(x) > 0 \text{ and } \sigma'(x) \to 0 \text{ as } x \to \pm\infty,\\
&\text{(iv)} && \sigma'(x) \text{ has a global maximum equal to 1.}
\end{aligned} \tag{6.14}$$

The activation function $\sigma(x)$ is a short map, hence it is obvious that

$$|\sigma(x)| \le |x|. \tag{6.15}$$

Lemma 6.1 *The Lipschitz constant of the system represented by (6.13) is:*

$$L = \max\{\|\boldsymbol{V}_2\boldsymbol{V}_1^x\|, \|\boldsymbol{V}_2\boldsymbol{V}_1^u\|\}$$

Proof Taking into account the form of the model (6.13) and the property of the activation function (6.15), it follows that:

$$\begin{aligned}
&\|\hat{g}(\boldsymbol{x}_1,\boldsymbol{u}_1) - \hat{g}(\boldsymbol{x}_2,\boldsymbol{u}_2)\|\\
&\quad \le \|\boldsymbol{V}_2\sigma(\boldsymbol{V}_1^x\boldsymbol{x}_1 + \boldsymbol{V}_1^u\boldsymbol{u}_1 + \boldsymbol{\beta}_1) + \boldsymbol{\beta}_2\\
&\qquad - \boldsymbol{V}_2\sigma(\boldsymbol{V}_1^x\boldsymbol{x}_2 + \boldsymbol{V}_1^u\boldsymbol{u}_2 + \boldsymbol{\beta}_1) - \boldsymbol{\beta}_2\|\\
&\quad \le \|\boldsymbol{V}_2(\boldsymbol{V}_1^x\boldsymbol{x}_1 + \boldsymbol{V}_1^u\boldsymbol{u}_1 + \boldsymbol{\beta}_1)\\
&\qquad - \boldsymbol{V}_2(\boldsymbol{V}_1^x\boldsymbol{x}_2 + \boldsymbol{V}_1^u\boldsymbol{u}_2 + \boldsymbol{\beta}_1)\|\\
&\quad \le \|\boldsymbol{V}_2\boldsymbol{V}_1^x(\boldsymbol{x}_1 - \boldsymbol{x}_2) + \boldsymbol{V}_2\boldsymbol{V}_1^u(\boldsymbol{u}_1 - \boldsymbol{u}_2)\|\\
&\quad \le \|\boldsymbol{V}_2\boldsymbol{V}_1^x\|\|\boldsymbol{x}_1 - \boldsymbol{x}_2\| + \|\boldsymbol{V}_2\boldsymbol{V}_1^u\|\|\boldsymbol{u}_1 - \boldsymbol{u}_2\|.
\end{aligned} \tag{6.16}$$

Let

$$L = \max\{\|\boldsymbol{V}_2\boldsymbol{V}_1^x\|, \|\boldsymbol{V}_2\boldsymbol{V}_1^u\|\} \tag{6.17}$$

then

$$\|g(\boldsymbol{x}_1,\boldsymbol{u}_1) - g(\boldsymbol{x}_2,\boldsymbol{u}_2)\| \le L(\|\boldsymbol{x}_1 - \boldsymbol{x}_2\| + \|\boldsymbol{u}_1 - \boldsymbol{u}_2\|).$$

6.3.2 Neural Controller

Let us consider the learning controller in the form:

$$u_p^{ff}(k) = f(\varphi_{p-1}(k)), \tag{6.18}$$

where $f(\cdot)$ is a nonlinear function, $\varphi_{p-1}(k)$ is a regression vector from the previous trial. The form of the regression vector determines the controller structure:

- for P-type controller:

$$\varphi_{p-1}(k) = [u_{p-1}(k), e_{p-1}(k)]^{\mathsf{T}}, \tag{6.19}$$

- for D-type controller:

$$\varphi_{p-1}(k) = [u_{p-1}(k), e_{p-1}(k+1)]^{\mathsf{T}}, \tag{6.20}$$

- for PD-type controller:

$$\varphi_{p-1}(k) = [u_{p-1}(k), e_{p-1}(k), e_{p-1}(k+1)]^{\mathsf{T}}. \tag{6.21}$$

Such a neural-controller structure is very flexible, since an application of different types of feedback data is possible by introducing relevant changes in the regressor vector only. The purpose of a controller synthesis is to find a nonlinear function $f(\cdot)$ in such a way as to achieve a possibly minimal tracking error $e_p(k)$ during the pth trial. Taking into account the fact that the mathematical model of a nonlinear plant can be hard or even impossible to derive, the ILC controller synthesis relies on a suitably designed neural model of a plant. In such a case, the proposed ILC scheme uses two neural networks: the first one plays the role of a model of the plant, while the second is the learning controller. Let us consider a neural network controller with one hidden layer:

$$u_p^{ff}(k) = \boldsymbol{W}_{2,p}\sigma(\boldsymbol{W}_{1,p}\varphi_{p-1}(k) + \boldsymbol{b}_{1,p}) + \boldsymbol{b}_{2,p}, \tag{6.22}$$

where $\boldsymbol{W}_{1,p} \in \mathbb{R}^{v_c \times 2}$ and $\boldsymbol{W}_{2,p} \in \mathbb{R}^{1 \times v_c}$ are weight matrices of hidden and output layers at pth trial, respectively, $\boldsymbol{b}_{1,p} \in \mathbb{R}^{v_c}$ and $\boldsymbol{b}_{2,p} \in \mathbb{R}^1$ are bias vectors of hidden and output units, respectively, $\sigma : \mathbb{R}^{v_c} \to \mathbb{R}^{v_c}$ is the activation function of the hidden layer and v_c stands for the number of hidden neurons. The neural network (6.22) can be viewed as a general representation of the ILC controller of the appropriate kind depending on the selection of the regression vector $\varphi_{p-1}(k)$ recorded during the previous trial.

6.3.3 An Update Rule

The idea is to update neural network controller parameters after each trial according to the rule:

$$\boldsymbol{\theta}_p = \boldsymbol{\theta}_{p-1} + \Delta\boldsymbol{\theta}_p, \tag{6.23}$$

where $\boldsymbol{\theta}_p$ stands for the generalized network parameter vector during the pth trial. The vector $\boldsymbol{\theta}_p \in \mathbb{R}^M$ represents all unknown network parameters, i.e. $\boldsymbol{\theta}_p = [\mathrm{svec}(\boldsymbol{W}_{1,p})^{\mathsf{T}}, \boldsymbol{W}_{2,p}, \boldsymbol{b}_{1,p}^{\mathsf{T}}, \boldsymbol{b}_{2,p}]^{\mathsf{T}}$, $\Delta\boldsymbol{\theta}_p$ is a correction term, and $M = 4v_c + 1$ is the number of the neural network controller parameters. The objective of the training is to determine the network parameters in such a way as to minimize the cost function:

$$\boldsymbol{\theta}_p^{\star} = \arg\min J_p, \tag{6.24}$$

where $\boldsymbol{\theta}_p^{\star}$ is the optimal controller parameter vector at the pth trial. Let us define the following cost function:

$$J_p = \frac{1}{2}\sum_{k=1}^{N}(y_r(k) - y_p(k))^2 + \frac{1}{2}\mu\sum_{i=1}^{M}\theta_{p,i}^2, \tag{6.25}$$

where μ is a parameter governing how strongly large weights are penalized, and $\theta_{p,i}$ is the ith component of the vector $\boldsymbol{\theta}_p$. The second term in (6.25) is a weight decay, which is a simple way of achieving an appropriate generalization of the neural network controller. Such a technique can prevent large changes of the weight values and thus the neural controller will not forget the training patterns presented during the previous trials.

The simplest and non-time consuming solution is to apply the gradient descent to minimize the cost (6.25) with respect to the parameter vector $\boldsymbol{\theta}_p$. The following weight update is derived:

$$\Delta\boldsymbol{\theta}_p = -\eta\frac{\partial J_p}{\partial\boldsymbol{\theta}_p}, \tag{6.26}$$

where η represents the learning rate. The gradient of the cost function (6.25) with respect to the vector $\boldsymbol{\theta}_p$ can be derived using the chain rule of differentiation as follows [20, 21]:

$$\begin{aligned}\frac{\partial J_p}{\partial\boldsymbol{\theta}_p} &= -\sum_{k=1}^{N}\left(e_p(k)\frac{\partial y_p(k)}{\partial u_p(k-1)}\frac{\partial u_p(k-1)}{\partial\boldsymbol{\theta}_p}\right) + \lambda\boldsymbol{\theta}_p \\ &= -\sum_{k=1}^{N}\left(e_p(k)\frac{\partial\hat{y}_p(k)}{\partial u_p(k-1)}\frac{\partial u_p(k-1)}{\partial\boldsymbol{\theta}_p}\right) + \lambda\boldsymbol{\theta}_p\end{aligned}. \tag{6.27}$$

Here we apply the equivalence rule saying that a model represents the plant sufficiently well, i.e. $y_p(k) \approx \hat{y}_p(k)$. That is the reason why the neural model is included in the scheme shown in Fig. 6.3. The first partial derivative in (6.27) can be determined using (6.12)–(6.13). We have:

$$\frac{\partial \hat{y}_p(k)}{\partial u_p(k-1)} = \boldsymbol{C}\boldsymbol{V}_2 \left(\sigma' \circ \boldsymbol{V}_1^u\right), \tag{6.28}$$

where $\boldsymbol{V}_1^u$ is the weight vector associated with the input $u_p(k-1)$, σ' stands for the derivative of the activation function and $\circ$ represents the Hadamard product (an element-wise product). The second partial derivative in (6.27) can be calculated using the neural controller (6.18). For simplicity let consider that the feedback controller gives a response independent on the learning controller parameters (or equivalently let us assume that $u_p^{fb}(k)$ is a constant value). Then:

- for weights of the first layer $\boldsymbol{W}_{1,p}$:

$$\frac{\partial u_p(k-1)}{\partial \boldsymbol{W}_{1,p}} = \left(\boldsymbol{W}_{2,p} \circ \sigma'\right) \boldsymbol{\varphi}_{p-1}(k-1), \tag{6.29}$$

- for biases of the first layer $\boldsymbol{b}_{1,p}$:

$$\frac{\partial u_p(k-1)}{\partial \boldsymbol{b}_{1,p}} = \boldsymbol{W}_{2,p} \circ \sigma', \tag{6.30}$$

- for weights of the second layer $\boldsymbol{W}_{2,p}$:

$$\frac{\partial u_p(k-1)}{\partial \boldsymbol{W}_{2,p}} = \sigma(\boldsymbol{W}_{1,p}\boldsymbol{\varphi}_{p-1}(k-1) + \boldsymbol{b}_{1,p}), \tag{6.31}$$

- for biases of the second layer $\boldsymbol{b}_{2,p}$:

$$\frac{\partial u_p(k-1)}{\partial \boldsymbol{b}_{2,p}} = \mathbf{1}. \tag{6.32}$$

6.4 Convergence Analysis

The ILC problem can be formulated as the problem of finding such a controller structure ensuring that the following property is satisfied:

P1: Convergence

$$\lim_{p\to\infty} u_p(k) = u^*(k), \tag{6.33}$$

First, let us start with an analysis of a convergence of the ILC scheme with a neural network controller implementing the P-type rule. We introduce the following notions:

Definition 6.2 The partial derivatives of the controller (6.18) and (6.19) with respect to the inputs are defined as follows:

$$f_u(k) = \frac{\partial f(u_p(k), e_p(k))}{\partial u_p(k)}, \quad f_e(k) = \frac{\partial f(u_p(k), e_p(k))}{\partial e_p(k)}. \tag{6.34}$$

Definition 6.3 Let the λ-norm of a vector $z(k)$ is defined as follows:

$$\|z(k)\|_\lambda = \sup_{k \in \mathbb{N}_{<N}} \alpha^{-\lambda k} \|z(k)\|, \tag{6.35}$$

where $\lambda > 0$ have to be a finite constant, and $\alpha > 1$ should be suitably selected.

Now, let us formulate the sufficient condition for the convergence of a P-type neural controller.

Theorem 6.4 *For the nonlinear system (6.8) under the Assumptions A1–A3, the convergence of the control law (6.18) with the regression vector (6.19) is guaranteed, if*

$$\gamma_1 + \gamma_2 \cdot \left(\frac{1-\alpha^{-(\lambda-1)N}}{1-\alpha^{-(\lambda-1)}} - 1 \right) < 1, \tag{6.36}$$

where $\gamma_1 = \sup_k \| f_u(k) \|$ *and* $\gamma_2 = \sup_k \| f_e(k) \boldsymbol{C} \|$.

Proof The argument can be derived via a generalization of the approach in [25]. First, applying Taylor's theorem to the control law (6.18) with the regression vector (6.19) and taking a first-order approximation we obtain:

$$\Delta u_{p+1}(k) = f_u(k) \Delta u_p(k) - f_e(k) \Delta y_p(k), \tag{6.37}$$

where $\Delta u_p(k) = u_r(k) - u_p(k)$ represents the input error, and $e_p(k) = \Delta y_p(k) = y_r(k) - y_p(k)$ stands for the tracking output error. For the system (6.8) from Assumption A1 it follows that

$$\begin{aligned} \Delta \boldsymbol{x}_p(k+1) &= g(\boldsymbol{x}_r(k), u_r(k)) - g(\boldsymbol{x}_p(k), u_p(k)) \\ \Delta y_p(k) &= \boldsymbol{C} \Delta \boldsymbol{x}_p(k) \end{aligned}, \tag{6.38}$$

where $\Delta \boldsymbol{x}_p(k) = \boldsymbol{x}_r(k) - \boldsymbol{x}_p(k)$. Substituting (6.38) into (6.37) yields:

$$\Delta u_{p+1}(k) = f_u(k) \Delta u_p(k) - f_e(k) \boldsymbol{C} \Delta \boldsymbol{x}_p(k). \tag{6.39}$$

Taking the norm of the both sides of (6.39) we obtain:

$$\begin{aligned}\|\Delta u_{p+1}(k)\| &= \|f_u(k)\Delta u_p(k) - f_e(k)\boldsymbol{C}\Delta \boldsymbol{x}_p(k)\| \\ &\leq \|f_u(k)\Delta u_p(k)\| + \|f_e(k)\boldsymbol{C}\Delta \boldsymbol{x}_p(k)\|.\end{aligned} \tag{6.40}$$

Now, taking the norm of the both sides of the state equation (6.38) and applying Assumption A3 we have:

$$\begin{aligned}\|\Delta \boldsymbol{x}_p(k+1)\| &= \|g(\boldsymbol{x}_r(k), u_r(k)) - g(\boldsymbol{x}_p(k), u_p(k))\| \\ &\leq \Big[L\|\boldsymbol{x}_r(k) - \boldsymbol{x}_p(k)\| + L|u_r(k) - u_p(k)|\Big] \\ &\leq L\|\Delta \boldsymbol{x}_p(k)\| + L|\Delta u_p(k)|.\end{aligned} \tag{6.41}$$

Taking into account the recursive nature of the system (6.41) and assuming that (A2) holds, we can write $\|\Delta \boldsymbol{x}_p(0)\| = 0$ and

$$\|\Delta \boldsymbol{x}_p(k)\| \leq \sum_{i=0}^{k-1} L^{k-i}|\Delta u_p(i)|, \quad k = 1, \ldots, N-1. \tag{6.42}$$

For the sake of simplifying notation, let us augment the sequence $\Delta u_p(k)$ with an initial zero element, i.e. let us introduce the following sequence $\widetilde{u}_p(k)$:

$$\widetilde{u}_p(k) = \begin{cases} 0, & k = 0, \\ |\Delta u_p(k-1)|, & k = 1, \ldots, N. \end{cases} \tag{6.43}$$

Then, substituting (6.42) into (6.40), we get for any $k \in \mathbb{N}_{<N}$:

$$\|\Delta u_{p+1}(k)\| \leq \|f_u(k)\Delta u_p(k)\| + \|f_e(k)\boldsymbol{C}\sum_{i=0}^{k} L^{k-(i-1)}\widetilde{u}_p(i)\|. \tag{6.44}$$

Applying λ-norm to the both sides of the inequality (6.44) yields:

$$\begin{aligned}&\sup_k \alpha^{-\lambda k}\|\Delta u_{p+1}(k)\| \\ &\qquad \leq \sup_k \alpha^{-\lambda k}\|f_u(k)\Delta u_p(k)\| \\ &\qquad\quad + \sup_k \alpha^{-\lambda k}\|f_e(k)\boldsymbol{C}\sum_{i=0}^{k} L^{k-(i-1)}\widetilde{u}_p(i)\| \\ &\qquad \leq \sup_k \alpha^{-\lambda k}\|f_u(k)\Delta u_p(k)\| \\ &\qquad\quad + \sup_k \|f_e(k)\boldsymbol{C}\| \sup_k \alpha^{-\lambda k}\sum_{i=0}^{k} L^{k-(i-1)}\widetilde{u}_p(i).\end{aligned} \tag{6.45}$$

Setting $L = \alpha$ and using (6.43), the last supremum in (6.45) satisfies:

$$
\begin{aligned}
&\sup_k \alpha^{-\lambda k} \sum_{i=0}^{k} \alpha^{k-i+1} \widetilde{u}_p(i) \\
&= \sup_k \sum_{i=0}^{k} \alpha^{-\lambda k+(k-i+1)} \widetilde{u}_p(i) \\
&= \sup_k \sum_{i=0}^{k} \alpha^{-\lambda(i-1)} \widetilde{u}_p(i) \alpha^{-(\lambda-1)(k-i+1)} \\
&\leq \sup_k \left(\sum_{i=0}^{k} \sup_k \left(\alpha^{-\lambda k} \widetilde{u}_p(k+1) \right) \alpha^{-(\lambda-1)(k-i+1)} \right) \\
&= \sup_k \alpha^{-\lambda k} |\Delta u_p(k)| \cdot \sup_k \sum_{i=0}^{k} \alpha^{-(\lambda-1)(k-i+1)} \\
&= \|\Delta u_p(k)\|_\lambda \cdot \alpha^{-(\lambda-1)} \cdot \frac{1-\alpha^{-(\lambda-1)N}}{1-\alpha^{-(\lambda-1)}}.
\end{aligned}
\tag{6.46}
$$

Finally, (6.45) and (6.46) leads to

$$
\begin{aligned}
\|\Delta u_{p+1}(k)\|_\lambda &\leq \sup_k \|f_u(k)\| \|\Delta u_p(k)\|_\lambda \\
&\quad + \sup_k \|f_e(k) \boldsymbol{C}\| \|\Delta u_p(k)\|_\lambda \left(\frac{1-\alpha^{-(\lambda-1)N}}{1-\alpha^{-(\lambda-1)}} - 1 \right) \\
&\leq \left(\gamma_1 + \gamma_2 \left(\frac{1-\alpha^{-(\lambda-1)N}}{1-\alpha^{-(\lambda-1)}} - 1 \right) \right) \|\Delta u_p(k)\|_\lambda,
\end{aligned}
\tag{6.47}
$$

where $\gamma_1 = \sup_k \|f_u(k)\|$ and $\gamma_2 = \sup_k \|f_e(k)\boldsymbol{C}\|$. It is clear that if

$$
\gamma_1 + \gamma_2 \left(\frac{1-\alpha^{-(\lambda-1)N}}{1-\alpha^{-(\lambda-1)}} - 1 \right) < 1
$$

then

$$
\forall k \in \mathbb{N}_{<N} \quad \lim_{p \to \infty} \|\Delta u_p(k)\|_\lambda = 0, \tag{6.48}
$$

which implies that

$$
\forall k \in \mathbb{N}_{<N} \quad \lim_{p \to \infty} \|\Delta u_p(k)\| = 0. \tag{6.49}
$$

Finally, from (6.49) it can be deduced that $\forall k \in \mathbb{N}_{<N} \ \lim_{p \to \infty} u_p(k) = u_r(k)$ which means that the convergence property (P1) is satisfied. □

Remark 6.5 Theorem 6.4 requires that the Lipschitz constant have to be greater than 1. In the view of Lemma 6.1, we know that L is strongly dependent on the neural model weight matrices V_2, V_1^u and V_1^x. Therefore. during the model training, it should be checked at each training epoch if the following condition is satisfied:

$$\max\{|V_2 V_1^x|, |V_2 V_1^u|\} > 1. \tag{6.50}$$

As the matrix V_2 is common in both arguments of the maximum operator, a very simple heuristic procedure can be proposed as described in Algorithm 6.1.

Algorithm 6.1 Lipschitz constant checking

Step 0. Set $\mu > 1$.

Step 1. Perform the weight update.

Step 2. Check the condition (6.50).

Step 3. **If** the condition (6.50) is not satisfied **then**

$$V_2^{new} = \mu V_2^{old}$$

and go to Step 1 **else** go to Step 4.

Step 4. End of training

Remark 6.6 A careful analysis of the condition (6.36) leads to an interesting observation that the convergence of the ILC scheme with the neural controller and a P-type update depends far more on the controller sensitivity with respect to the control input from the previous trial than on the error-input sensitivity. In fact, if it is assumed that $\alpha > 1$ then

$$\lim_{\lambda\to\infty} \alpha^{-(\lambda-1)} = 0, \tag{6.51}$$

and similarly, as N is positive

$$\lim_{\lambda\to\infty} \alpha^{-N(\lambda-1)} = 0. \tag{6.52}$$

Hence, we get:

$$\lim_{\lambda\to\infty} \frac{1-\alpha^{-(\lambda-1)N}}{1-\alpha^{-(\lambda-1)}} - 1 = 0, \tag{6.53}$$

and the condition (6.36) reduces to:

$$\gamma_1 < 1. \tag{6.54}$$

So we can focus only on the magnitude of the derivative $f_u(\cdot)$, treating $f_e(\cdot)$ as an additional safety margin ensuring convergence.

As for practical implementation of the condition resulting from Theorem 6.4, consider the controller representation (6.22) and (6.19). Then, the partial derivatives (6.34) take the form:

$$f_u(k) = \left(\boldsymbol{W}_{2,p} \circ \sigma'\right) \boldsymbol{W}^u_{1,p}, \quad f_e(k) = \left(\boldsymbol{W}_{2,p} \circ \sigma'\right) \boldsymbol{W}^e_{1,p}, \tag{6.55}$$

where $\boldsymbol{W}^u_{1,p} \in \mathbb{R}^1$ and $\boldsymbol{W}^e_{1,p}\mathbb{R}^1$ are weights associated with the input $u_p(k)$ and tracking error $e_p(k)$, respectively. According to the condition (6.14iv), the supremum of a derivative of the activation function is equal to one. Moreover, if we consider SISO systems, from the structure of the neural network model we know that the output matrix is in the form of the unit vector ($\boldsymbol{C} = [1, 0, \ldots, 0]$). Then

$$\gamma_1 = \|\boldsymbol{W}_{2,p}\boldsymbol{W}^u_{1,p}\|, \quad \gamma_2 = \|\boldsymbol{W}_{2,p}\boldsymbol{W}^e_{1,p}\| \tag{6.56}$$

and the condition (6.54) can be rewritten as follows:

$$\|\boldsymbol{W}_{2,p}\boldsymbol{W}^u_{1,p}\| + \|\boldsymbol{W}_{2,p}\boldsymbol{W}^e_{1,p}\| < 1, \tag{6.57}$$

Now, let us investigate the convergence properties of ILC with a neural network controller of the D-type. By analogy to (6.34), let us start with the definition of the sensitivities of a controller:

Definition 6.7 The partial derivatives of the controller (6.18) with respect to the components of regression vector (6.20) are defined as:

$$f^{\mathrm{D}}_u(k) = \frac{\partial f(u_p(k), e_p(k+1))}{\partial u_p(k)}, \quad f^{\mathrm{D}}_e(k) = \frac{\partial f(u_p(k), e_p(k+1))}{\partial e_p(k+1)}. \tag{6.58}$$

Theorem 6.8 *For the nonlinear system (6.8), under the Assumptions A1–A3, the convergence of the control law (6.18) with the regression vector (6.20) is guaranteed, if*

$$\gamma_3 + \gamma_4 \cdot \alpha \frac{1 - \alpha^{-(\lambda-1)N}}{1 - \alpha^{-(\lambda-1)}} < 1, \tag{6.59}$$

where $\gamma_3 = \sup_k \|f^{\mathrm{D}}_u(k)\|$ *and* $\gamma_4 = \sup_k \|f^{\mathrm{D}}_e(k)\boldsymbol{C}\|$.

Proof The reasoning is somewhat similar to the P-type controller case. Applying Taylor's theorem to the control law (6.18) with the regression vector (6.20) and a first-order approximation we get:

$$\Delta u_{p+1}(k) = f^{\mathrm{D}}_u(k)\Delta u_p(k) - f^{\mathrm{D}}_e(k)\Delta y_p(k+1). \tag{6.60}$$

Substituting (6.38) into (6.60) gives:

$$\Delta u_{p+1}(k) = f^{\mathrm{D}}_u(k)\Delta u_p(k) - f^{\mathrm{D}}_e(k)\boldsymbol{C}\Delta \boldsymbol{x}_p(k+1). \tag{6.61}$$

Taking the norm of the both sides of (6.61) we obtain:

$$\|\Delta u_{p+1}(k)\| \le \|f^{\mathrm{D}}_u(k)\Delta u_p(k)\| + \|f^{\mathrm{D}}_e(k)\boldsymbol{C}\Delta \boldsymbol{x}_p(k+1)\|. \tag{6.62}$$

Analyzing the recursive nature of the system (6.41) and taking into account the Assumption A2 we can write:

$$\|\Delta \boldsymbol{x}_p(k+1)\| \leq \sum_{i=0}^{k} L^{k-i+1} |\Delta u_p(i)|. \tag{6.63}$$

Substituting (6.63) into (6.62) gives:

$$\begin{aligned} \|\Delta u_{p+1}(k)\| &\leq \|f_u^{\mathrm{D}}(k) \Delta u_p(k)\| \\ &+ \|f_e^{\mathrm{D}}(k) \boldsymbol{C} \sum_{i=0}^{k} L^{k-i+1} |\Delta u_p(i)|\| \end{aligned} \tag{6.64}$$

Let us apply λ-norm to the both sides of the inequality (6.64), then

$$\begin{aligned} &\|\Delta u_{p+1}(k)\|_\lambda \\ &\leq \sup_k \|f_u^{\mathrm{D}}(k)\| \|\Delta u_p(k)\|_\lambda \\ &\quad + \sup_k \|f_e^{\mathrm{D}}(k) \boldsymbol{C}\| \sup_k \alpha^{-\lambda k} \sum_{i=0}^{k} L^{k-i+1} |\Delta u_p(i)|. \end{aligned} \tag{6.65}$$

Setting $L = \alpha$ and performing calculations analogously to (6.45), the last supremum in (6.65) is going to be:

$$\begin{aligned} &\sup_k \alpha^{-\lambda k} \sum_{i=0}^{k} \alpha^{k-i+1} |\Delta u_p(i)| \\ &= \|\Delta u_p(k)\|_\lambda \cdot \alpha \frac{1 - \alpha^{-(\lambda-1)N}}{1 - \alpha^{-(\lambda-1)}}. \end{aligned} \tag{6.66}$$

Thus, we have:

$$\|\Delta u_{p+1}(k)\|_\lambda \leq \left(\gamma_3 + \gamma_4 \cdot \alpha \frac{1-\alpha^{-(\lambda-1)N}}{1-\alpha^{-(\lambda-1)}} \right) \|\Delta u_p(k)\|_\lambda. \tag{6.67}$$

It is clear, that if only

$$\gamma_3 + \gamma_4 \cdot \alpha \frac{1-\alpha^{-(\lambda-1)N}}{1-\alpha^{-(\lambda-1)}} < 1$$

then

$$\forall k \in \mathbb{N}_{<N} \quad \lim_{p \to \infty} \|\Delta u_p(k)\|_\lambda = 0 \tag{6.68}$$

which implies that

$$\forall k \in \mathbb{N}_{<N} \quad \lim_{p\to\infty} \|\Delta u_p(k)\| = 0. \tag{6.69}$$

Then, based on (6.69) it can be deduced that

$$\forall\, k \in \mathbb{N}_{<N} \quad \lim_{p\to\infty} \Delta u_p(k) = 0,$$

which means that the convergence property (P1) is satisfied. □

Remark 6.9 According to Remark 6.6, for sufficiently large λ the condition (6.59) can be approximated as follows:

$$\gamma_3 + \gamma_4\alpha < 1. \tag{6.70}$$

Since the derivatives (6.58) have the same representation as (6.55), the condition (6.70) takes the form:

$$\|\boldsymbol{W}_{2,p}\boldsymbol{W}^u_{1,p}\| + L\|\boldsymbol{W}_{2,p}\boldsymbol{W}^e_{1,p}\| < 1. \tag{6.71}$$

6.5 An Experimental Study — The Pneumatic Servo Revisited

Let us revisit the pneumatic servomechanism described in detail in Chap. 4, Sect. 4.6.

6.5.1 Modelling

The first step of the control synthesis is to design a model of the plant. For the reasons presented in Sect. 4.6, the data were recorded in a closed-loop control. The training set $\{u_p(k), y_p(k)\}_{k=1}^N$ consists of 16000 samples. The order of the model (6.8) was selected after analyzing the physical properties of the process. The best settings for neural network model were as follows: $n_x = 3$ and $v_m = 5$. Such settings ensure acceptable modelling results. Hidden neurons consist of hyperbolic tangent activation function, while the output neuron has a linear activation function. The training was carried out off-line for 100 steps with the Levenberg–Marquardt algorithm. The obtained quality index $\mathrm{SSE}(y, \hat{y}) = 0.0438$ guarantees a decent level of modelling performance.

6.5.2 ILC Synthesis

In order to satisfy the convergence condition (6.36), the following simple heuristic can be used. First, one needs to introduce a safety margin (in our case it was set to 0.99). Then, if the condition (6.36) is greater than the safety level, the weights are not updated as described by the training algorithm but instead are set back to the best values saved during the learning phase. The best weights are defined as those parameters, for which the lowest value of the tracking error norm was achieved during the learning.

Initially, the parameters of the ILC controller (6.18)–(6.19) are chosen to be random values. The neural controller is retrained after each trial according to the algorithm presented in Sect. 6.3.3. In order to avoid over-fitting, the number of training epochs should be kept relatively small (only 50 in this case). Through a series of experiments, the value of the training rate was found to be $\eta = 0.05$ and the weight decay was set to $\lambda = 0.0001$ providing a good performance level of the learning process. Figure 6.4a shows the quality of the control over consecutive trials. At the beginning we observe a significant improvement but later on the error norm is increasing in spite of the fact that the convergence condition is satisfied (cf. Fig. 6.4c). This is a well-known phenomenon commonly observed in many machine-learning algorithms and is called a *learning transient*. In order to reduce this undesirable effect and to achieve the monotonic convergence, one simple and effective approach is to use a low pass Q-filter, which can disable learning at high frequencies. Thus, learning transients can be eliminated and, simultaneously, robustness of the control can be increased. The modified learning rule takes the form:

$$u_p^{ff}(k) = Q(z) f(\varphi_{p-1}(k)), \tag{6.72}$$

with the Q-filter described by the transfer function:

$$Q(z) = \frac{0.1912}{z - 0.8088}. \tag{6.73}$$

The cut-off frequency of the filter was set to $0.034 \frac{\text{rad}}{\text{s}}$ resulting in a very good effectiveness. In Fig. 6.4b, we can see the effect of introducing the control law (6.72) with the Q-filter (6.73) (blue line). This time, the learning transients are eliminated giving nearly monotonic convergence of the ILC scheme.

A very interesting observation is presented in Fig. 6.5, where the control signal (dashed) is plotted as well as its feedback (dash-dot) and feed-forward (solid) components. It is evident that during the learning process, the feed-forward neural controller from trial to trial gradually takes over and after approximately 50 iterations has the main impact on the control signal synthesis.

The tracking results are presented in Fig. 6.6. The reference trajectory is represented with the blue dash-dot line. Firstly, we investigated the system with the feedback P controller only (the solid line in Fig. 6.6a). In this case, the tracking error norm was $\|e_p(k)\| = 0.3156$. Clearly, the P controller has visible problems with an accurate tracking of the reference. Also, there are difficulties with achieving zero error at the steady-state. Therefore, in the second experiment, the classical control system with the P controller was enhanced by means of ILC using the scheme shown in Fig. 6.3. The system response in this case is depicted in Fig. 6.6b). In contrast to the P controller, the results achieved by ILC are definitely much improved. The

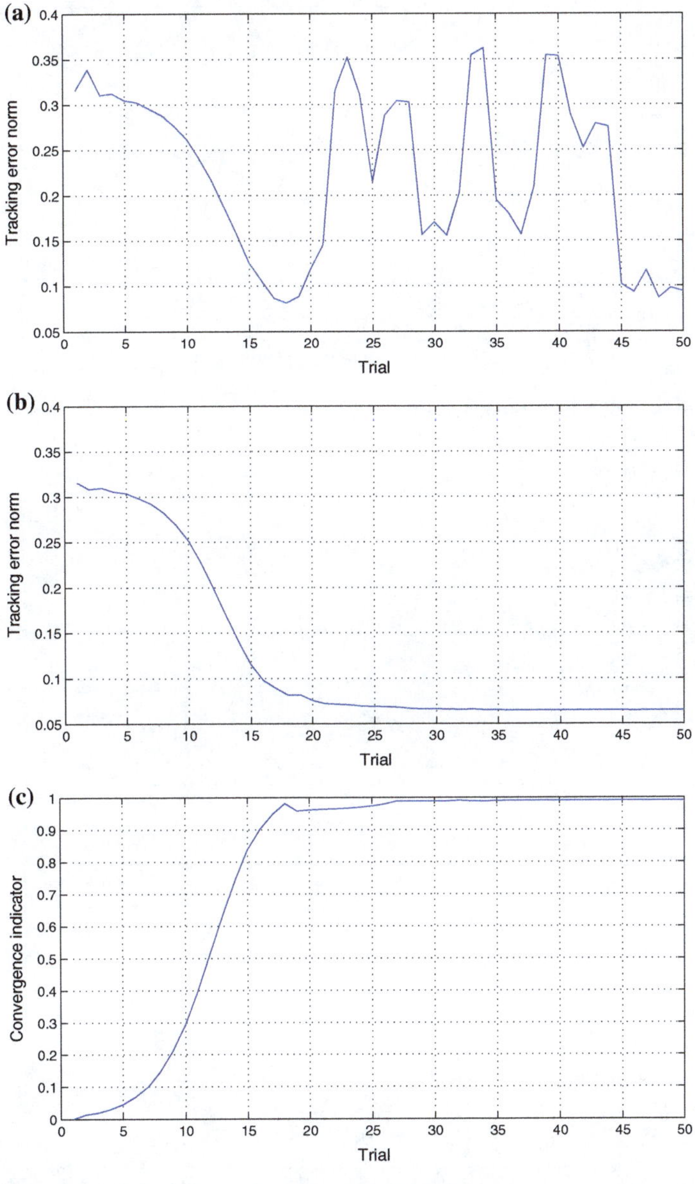

Fig. 6.4 A transient behaviour of ILC without filtering (**a**), with filtering (**b**) versus the plot of the convergence condition (**c**)

value of the tracking error norm was $\|e_p(k)\| = 0.0645$ in this case. What is also very important here, ILC is able to achieve a near zero steady-state error.

The convergence rate of the ILC with the P type update rule was compared to other learning controllers, namely D-type, cf. the regressor vector (6.20), and

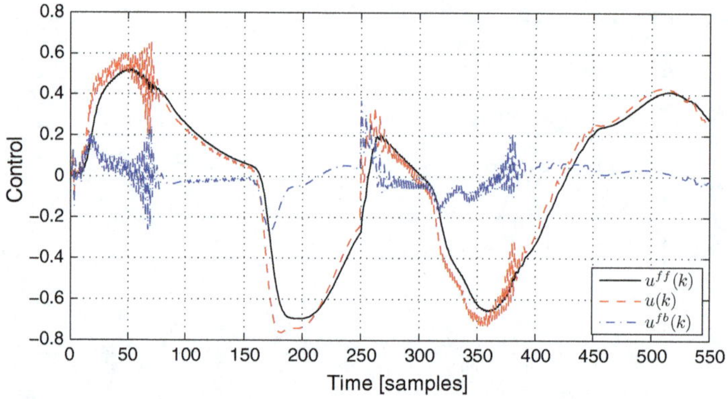

Fig. 6.5 The control signal components

(a)

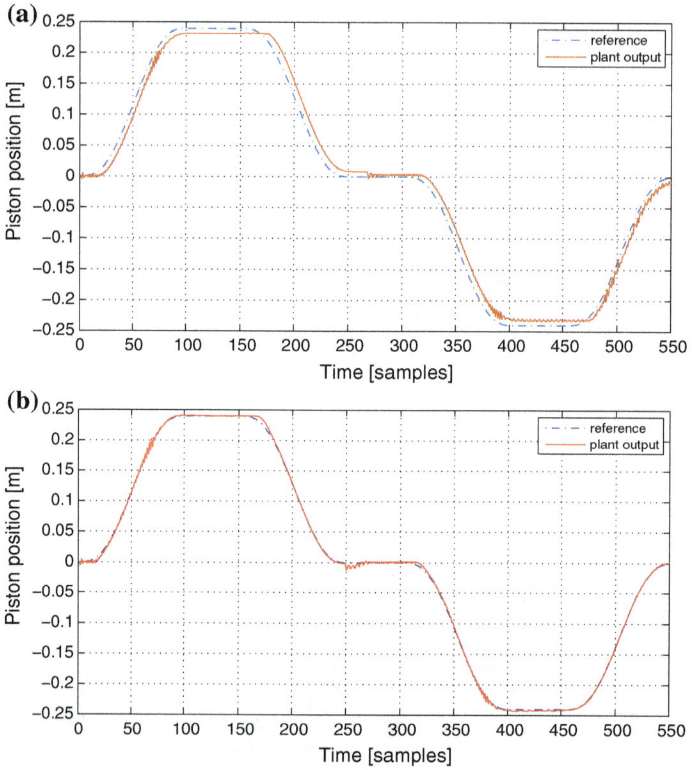

(b)

Fig. 6.6 Reference tracking: the reference (dash-dot) and the plant output (solid), the PID controller (**a**), and ILC (**b**)

PD-type (6.21). The results of the iterative learning are presented in Fig. 6.7, where ILC of the P-type is marked with the solid line, D-type with the dash-dot line and PD-type with the dashed line. The best results for the investigated process were

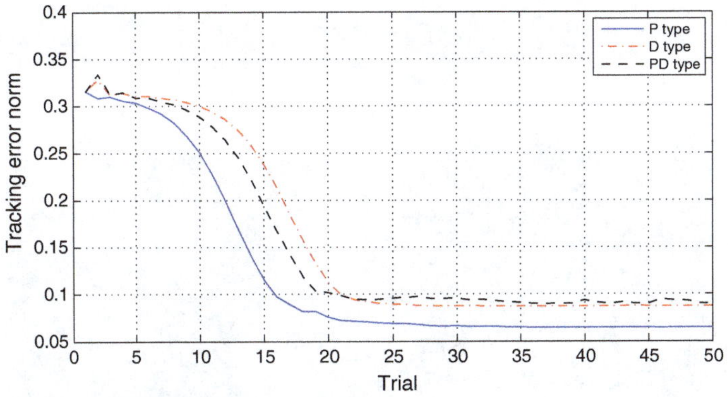

Fig. 6.7 A comparison of different ILC schemes

obtained using P-type controller. Both the convergence rate as well as the control quality are significantly better than achieved in cases of the other two controllers. It means that for the investigated system, the best choice is to apply a P-type ILC.

6.6 An Illustrative Example — A Magnetic Suspension System

Amira MA 401 Magnetic Suspension laboratory setup was used as an implementation of a nonlinear and unstable SISO system. The experimental stand consists of a metal body which hovers freely in the magnetic field, an electro-magnet (the actuator) and a comprehensive electronic control system. The equipment is shown in Fig. 6.8. The system is controlled using SPEEDGOAT real-time target machine, which allows the control design to be run, monitored and tested in real-time regime providing a quick path from MATLAB®/Simulink® interface to the experiments with the hardware plants.

The reference signal was set as the sinusoidal one with the frequency of 0.5 $\frac{\mathrm{rad}}{\mathrm{s}}$ and magnitude of 4. The sampling time is set to $T_s = 0.01$ s. The measured distance $y_p(k)$ varies from 0 to 5 mm, which corresponds to the voltage levels from -10 to 10 V. The applied control signal may vary from 0 to 10 V.

The first step was to design a feedback controller in order to stabilize the system. The proportional-integral-derivative controller was tuned using the pole placement by means of the linearized discrete-time model of the plant:

$$G(z) = \frac{0.2065z^2 + 0.5467z + 0.07699}{z^3 - 2.327z^2 + 1.297z - 0.1353} \tag{6.74}$$

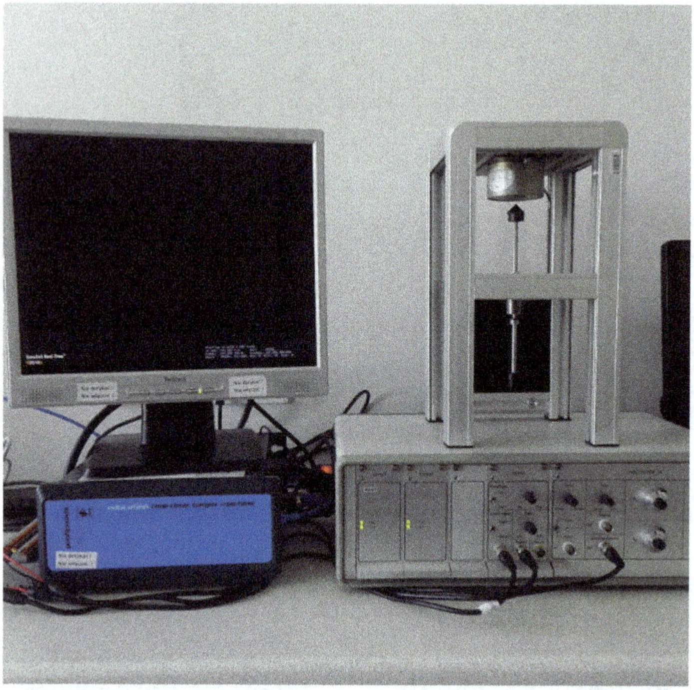

Fig. 6.8 The magnetic suspension laboratory stand

with the real poles: $z_1 = 0.1353$, $z_2 = 0.6474$ and $z_3 = 1.5446$. The plant is unstable as the pole z_3 is located outside the unit circle. The following discrete time transfer function of the PID controller was determined:

$$G_{PID}(z) = \frac{1.044z^2 - 1.716z + 0.69}{z^2 - z}. \tag{6.75}$$

The results of trajectory tracking for PID controller are presented in Fig. 6.9, where the reference is marked with the dotted line and the system output with the dashed line. The control problem presenting itself here is quite challenging, thus the controller encounters serious difficulties while following the reference. In addition to this, at the initial time instant, the PID controller requires some extra settling time in order to generate a control signal large enough to move the metal body to a required position.

For the modelling purposes, the data were collected in a closed control loop using a reference in the form of random steps. The measured input was the normalized electro-magnet voltage $u_p(k)$ and the modelled output was the distance of the metal body to the electro-magnet plate $y_p(k)$. In this example, the training dataset includes 20000 samples. After the 'trial and error' procedure, the best structure of the neural model (6.22) was determined as: $n_x = 3$ and $v_m = 5$. Again, hidden neurons consist of a hyperbolic tangent activation function, while the output neuron has the linear

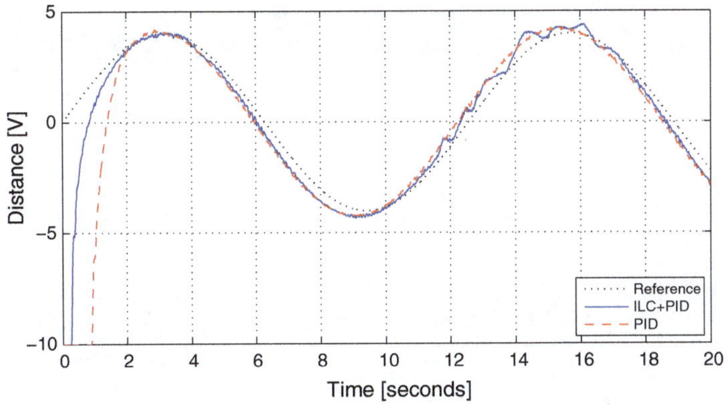

Fig. 6.9 The magnetic suspension: the trajectory tracking results

activation function. For the training purposes, and mainly to ensure stable learning, the input and output data were linearly scaled, dividing the signals by 10. Training was carried out off-line for 100 steps with the LM algorithm. After the training, the input weights were divided by 10 according to the input data scaling, in order to provide a proper operation during plant control. As the model of the plant is only used to calculate the sensitivities of the plant response with respect to the control signal based on the hidden neurons activation, the output scaling is not required in the implementation. Since the Lipschitz constant of the model is $L = 3.9144$, the requirements of Definition 6.3 are satisfied.

A P-type (6.18) learning controller with regressors (6.19) was used and, by analogy to the previous example, the initial weights of the controller were initialized with random values. After that, the neural controller was retrained in each trial in order to decrease the tracking error norm. The parameters of the training algorithm were set experimentally and the final values are: $\eta = 0.0005$ and $\lambda = 0.0001$. We also used the first-order Q-filter with the cut-off frequency equal to $7\,\frac{\text{rad}}{\text{s}}$ giving reasonable robustness of the learning controller. Figure 6.10 illustrates the convergence of the

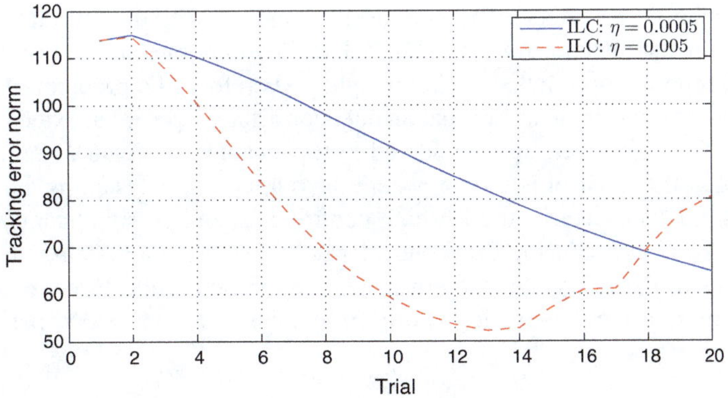

Fig. 6.10 The magnetic suspension: the tracking error norm

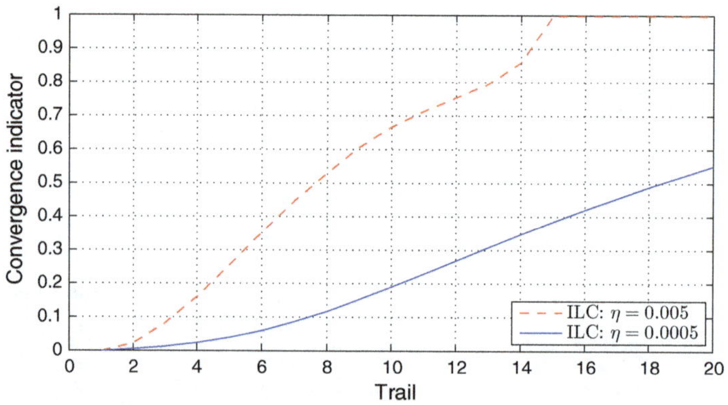

Fig. 6.11 The magnetic suspension: the convergence condition satisfaction

tracking error norm in the consecutive trials (the solid line). Taking into account the non-trivial dynamics of the plant, ILC performs on a decent level as the tracking norm is smoothly decreasing. In turn, Fig. 6.11 illustrates the evolution of the convergence condition. It is evident that the convergence indicator is satisfied assuring the convergence of ILC. However, a detailed analysis of trends observed on both figures leads to the observation that the tracking error norm can be further significantly decreased at the cost of a simultaneous increase of the value of the convergence indicator until it reaches the value very close to one. Thus, certain additional, yet important, issues appear here, which are related to a potential speed-up of the convergence process and a situation when the convergence condition obtains the upper allowable value. A possible solution is to increase the learning rate in order to faster train the neural controller (the case of increased learning rate $\eta = 0.005$ is plotted with the dashed line in Figs. 6.10 and 6.11). It is clear that the increased value of the learning rate is responsible for a much faster convergence of the control (see Fig. 6.10 dashed line up to 13 iteration). However, after that level, we are observing a deterioration of the control quality in spite of the fact that the convergence condition is satisfied (see Fig. 6.11 the dashed line). This phenomenon is closely connected with the learning process of the controller parameters. An exceedingly large learning rate causes problems with convergence to the optimal controller weights. Therefore, the problem of selecting the learning parameters in an optimal manner constitutes a potential extension of this work, which is, however, not considered in this chapter. In Fig. 6.9, the quality of the proposed ILC control is shown, where the output of the system is marked with the blue-solid line. Clearly, the learning controller is able to significantly improve the efficiency achieved using PID controller only. The control system reacts fast on the changes in the reference. Moreover, the learning controller tries to follow the reference more accurately, which is especially evident in the second period of the reference sine.

6.7 Dynamic Learning Controller

Let us consider the representation of the system in the form of the following difference equation:

$$y_p(k+1) = g(y_p(k), \ldots, y_p(k-n_a+1), u_p(k), \ldots, u_p(k-n_b+1), k), \tag{6.76}$$

where $g(\cdot)$ represents a nonlinear function which in general is unknown, $u_p(k)$ and $y_p(k)$ is input and output at the time instant k during the p trial, respectively, n_a and n_b are the number of delayed outputs and inputs, respectively, and N is the trial length.

The typical assumptions are that the plant is stable or satisfy some kind of Lipschitz conditions, the system returns to the same initial conditions at the start of each trial and each trial lasts for the same fixed time.

6.7.1 A Model of the System

A model of the system can be designed analogously to the case of the neural controller. Let consider a model of the nonlinear system (6.76) to be represented as:

$$y(k+1) = \hat{g}(\phi(k)), \tag{6.77}$$

where $\hat{g}$ is a nonlinear function, $\phi(k)$ is a regression vector of the form $\phi(k) = [y(k), \ldots, y(k-n_c+1), u(k), \ldots, u(k-n_d+1)]^\mathsf{T}$. Once again assume the neural model with one hidden layer:

$$\hat{y}(k+1) = \hat{f}(\phi(k)) = \sigma_o(\boldsymbol{V}_2^\mathsf{T}\sigma_h(\boldsymbol{V}_1^\mathsf{T}\varphi(k) + \boldsymbol{\beta}_1) + \boldsymbol{\beta}_2), \tag{6.78}$$

where $\boldsymbol{V}_1 \in \mathbb{R}^{(n_c+n_d)\times v_m}$ and $\boldsymbol{V}_2 \in \mathbb{R}^{v_m\times 1}$ are weight matrices of hidden and output layers, respectively, $\boldsymbol{\beta}_1 \in \mathbb{R}^{v_m}$ and $\boldsymbol{\beta}_2 \in \mathbb{R}^1$ are bias vectors of hidden and output units, respectively, $\sigma_h : \mathbb{R}^{v_m} \rightarrow \mathbb{R}^{v_m}$ is the activation function of the hidden layer, and $\sigma_o : \mathbb{R}^1 \rightarrow \mathbb{R}^1$ is the activation function of the output layer and v_m stands for the number of hidden neurons.

6.7.2 Neural Controller

The key idea here is very similar to the case of the static learning controller and is depicted in Fig. 6.3. We consider the control law (6.7) with the learning controller represented as [21]:

$$u_p^{ff}(k) = f(\varphi_{p-1}(k)), \tag{6.79}$$

where $f(\cdot)$ is a nonlinear function, $\varphi(k)$ is a regression vector of the form

$$\varphi_p(k) = [u_p(k), \ldots, u_p(k - n_c + 1), e_p(k), \ldots, e_p(k - n_d + 1)]^\mathsf{T}, \qquad (6.80)$$

where $e_p(k)$ is the tracking error at the time instant k during the p trial, n_c and n_d represents the number of past controls and tracking errors used by the controller. In order to estimate an unknown nonlinear mapping $f(\cdot)$, artificial neural networks can be used. Let us consider a neural model with one hidden layer:

$$u_p^{ff}(k) = \boldsymbol{W}_{2,p}\sigma(\boldsymbol{W}_{1,p}\boldsymbol{\varphi}_{p-1}(k) + \boldsymbol{b}_{1,p}) + \boldsymbol{b}_{2,p}, \qquad (6.81)$$

where $\boldsymbol{W}_{1,p} \in \mathbb{R}^{v_c \times (n_c+n_d)}$ and $\boldsymbol{W}_{2,p} \in \mathbb{R}^{1\times v_c}$ are weight matrices of hidden and output layers at pth trial, respectively, $\boldsymbol{b}_{1,p} \in \mathbb{R}^{v_c}$ and $\boldsymbol{b}_{2,p} \in \mathbb{R}^1$ are bias vectors of hidden and output units, respectively, $\sigma : \mathbb{R}^{v_c} \to \mathbb{R}^{v_c}$ is the activation function of the hidden layer and v_c stands for the number of hidden neurons. Finally, the regression vector is represented as (6.80). Such a neural model can be viewed as a NARX model discussed in Chap. 2, Sect. 2.3.1.

6.7.3 An Update Rule

In order to update the parameters of the dynamic controller, the algorithm introduced in Sect. 6.3.3 can be used. There are two modification, which should be taken into account. The first partial derivative in (6.27) can be determined using (6.78). It yields

$$\frac{\partial \hat{y}_p(k)}{\partial u_p(k-1)} = \boldsymbol{V}_2 \left(\sigma' \circ \boldsymbol{V}_1^u\right), \qquad (6.82)$$

where $\boldsymbol{V}_1^u$ is the weight vector associated with the input $u_p(k-1)$, σ' stands for the derivative of the activation function and $\circ$ represents the Hadamard product. The second partial derivative in (6.27) with respect to a specific kind of a neural controller parameter can be derived using rules (6.29)–(6.32), however, this time the regressor is represented by (6.80).

6.8 Stability Analysis

Considering the time domain, the neural network controller can be treated as a feed-forward network excited with the delayed controls and tracking errors recorded during the previous trial. Such a model is always stable because there is an algebraic relationship between a prediction of the control and the past measurements (control and tracking error). However, if this model is used in the framework of the iterative learning control, the crucial question that arises is related to the stability of the controller in the trial domain, i.e. the stability from trial to trial.

Let us transform the structure (6.81) into the state-space form. Let us define

$$\boldsymbol{x}_p = [u_p(k), \ldots, u_p(k - n_c + 1)]^{\mathsf{T}} \tag{6.83}$$

and

$$\boldsymbol{u}_p = [e_p(k), \ldots, e_p(k - n_d + 1)]^{\mathsf{T}} \tag{6.84}$$

as the state vector and input of the neural controller, respectively. Then, the state-space representation of the neural controller is:

$$\boldsymbol{x}_{p+1} = \boldsymbol{W}^x_{2,p}\sigma(\boldsymbol{W}^x_{1,p}\boldsymbol{x}_p + \boldsymbol{W}^u_{1,p}\boldsymbol{u}_p + \boldsymbol{b}_{1,p}) + \boldsymbol{b}^x_{2,p}, \tag{6.85}$$

where $\boldsymbol{W}^x_{2,p} \in \mathbb{R}^{n_c \times v_c}$ ($\boldsymbol{W}^x_{2,p} = [\boldsymbol{W}^{\mathsf{T}}_{2,p}\ \boldsymbol{0}^{\mathsf{T}}]^{\mathsf{T}}$), $\boldsymbol{b}^x_{2,p} \in \mathbb{R}^{n_c}$ ($\boldsymbol{b}^x_{2,p} = [\boldsymbol{b}^{\mathsf{T}}_{2,p}\ \boldsymbol{0}^{\mathsf{T}}]^{\mathsf{T}}$), $\boldsymbol{W}^x_{1,p} \in \mathbb{R}^{v_c \times n_c}$ and $\boldsymbol{W}^u_{1,p} \in \mathbb{R}^{v_c \times n_d}$ are weight matrices associated with the state and input vectors, respectively. Substituting $\boldsymbol{s}_1 = \boldsymbol{W}^u_{1,p}\boldsymbol{u}_p + \boldsymbol{b}_{1,p}$, where $\boldsymbol{s}_1$ can be regarded as a threshold or a fixed input and using the linear transformation $\boldsymbol{v}_p = \boldsymbol{W}^x_{1,p}\boldsymbol{x}_p + \boldsymbol{s}_1$ one obtains an equivalent system:

$$\boldsymbol{v}_{p+1} = \boldsymbol{W}^x_{1,p}\boldsymbol{W}^x_{2,p}\sigma(\boldsymbol{v}_p) + \boldsymbol{s}_2, \tag{6.86}$$

where $\boldsymbol{s}_2 = \boldsymbol{W}^x_{1,p}\boldsymbol{b}^x_{2,p} + \boldsymbol{s}_1$ is a threshold or a fixed input. Let $\boldsymbol{v}^*$ be an equilibrium point of (6.86), i.e. $\boldsymbol{v}^* = \boldsymbol{v}_{p+1} = \boldsymbol{v}_p$. Introducing an equivalent coordinate transformation $\boldsymbol{z}_p = \boldsymbol{v}_p - \boldsymbol{v}^*$, the system (6.86) can be transformed to the form:

$$\boldsymbol{z}_{p+1} = \boldsymbol{W}^x_{1,p}\boldsymbol{W}^x_{2,p}\bar{\sigma}(\boldsymbol{z}_p), \tag{6.87}$$

where

$$\bar{\sigma}(\boldsymbol{z}_p) = \sigma(\boldsymbol{z}_p + \boldsymbol{v}^*) - \sigma(\boldsymbol{v}^*). \tag{6.88}$$

Lemma 6.10 *The function $\bar{\sigma}(\boldsymbol{z}_p)$ defined in (6.88) satisfies the following condition:*

$$\|\bar{\sigma}(\boldsymbol{z}_p)\| \le \|\boldsymbol{z}_p\|. \tag{6.89}$$

Proof For the activation function $\sigma(\cdot)$ satisfying the conditions (6.14) and (6.15) the Lipschitz condition can be rewritten in the following form:

$$\|\sigma(\boldsymbol{z}_p + \boldsymbol{v}^*) - \sigma(\boldsymbol{v})\| \le \|(\boldsymbol{z}_p + \boldsymbol{v}^*) - \boldsymbol{v}^*\| = \|\boldsymbol{z}_p\|. \tag{6.90}$$

Substituting (6.88) into (6.90) we obtain:

$$\|\bar{\sigma}(\boldsymbol{z}_p)\| \le \|\boldsymbol{z}_p\|, \tag{6.91}$$

which completes the proof. □

Theorem 6.11 ([21]) *The neural system represented by (6.87) is globally asymptotically stable, if the following condition is satisfied:*

$$\|\boldsymbol{W}^x_{1,p}\boldsymbol{W}^x_{2,p}\| < 1. \tag{6.92}$$

Proof Let $V(z_p) = \|z_p\|$ be a Lyapunov function for the system (6.87). This function is positive definite with the minimum at $z_p = \mathbf{0}$. The difference along the trial trajectory is given as follows:

$$\Delta V(z_p) = \|z_{p+1} - z_p\| = \|\boldsymbol{W}^x_{1,p}\boldsymbol{W}^x_{2,p}\bar{\sigma}(z_p)\| - \|z_p\|. \tag{6.93}$$

According to Lemma 6.10, (6.93) can be expressed in the form:

$$\begin{aligned}\Delta V(z_p) &\leq \|\boldsymbol{W}^x_{1,p}\boldsymbol{W}^x_{2,p}z_p\| - \|z_p\| \\ &\leq (\|\boldsymbol{W}^x_{1,p}\boldsymbol{W}^x_{2,p}\| - 1)\|z_p\|.\end{aligned} \tag{6.94}$$

From (6.94) one can see that if

$$\|\boldsymbol{W}^x_{1,p}\boldsymbol{W}^x_{2,p}\| < 1, \tag{6.95}$$

then $\Delta V(z_p)$ is negative definite and the system (6.87) is globally asymptotically stable, which completes the proof. □

Remark 6.12 The criterion provided by Theorem 6.11 seems to be a restrictive one. In any case, it gives a very simple method of verification of the neural controller stability. This is an important result because assuming that the feedback controller is used mainly to stabilize the system, introducing additional stable control component will not destabilize the entire control system.

6.9 An Illustrative Example — The Pneumatic Servo Revisited

The pneumatic servomechanism was described in detail in Chap. 4, Sect. 4.6. The process model design was presented in Sect. 6.5.1.

6.9.1 Neural Controller

The neural-network controller was designed in such a way as to mimic the behaviour of the fundamental feedback controller. Thus, the input signal is the tracking error $e(k)$ and the desired output is the control signal $u(k)$. In this experiment, the neural

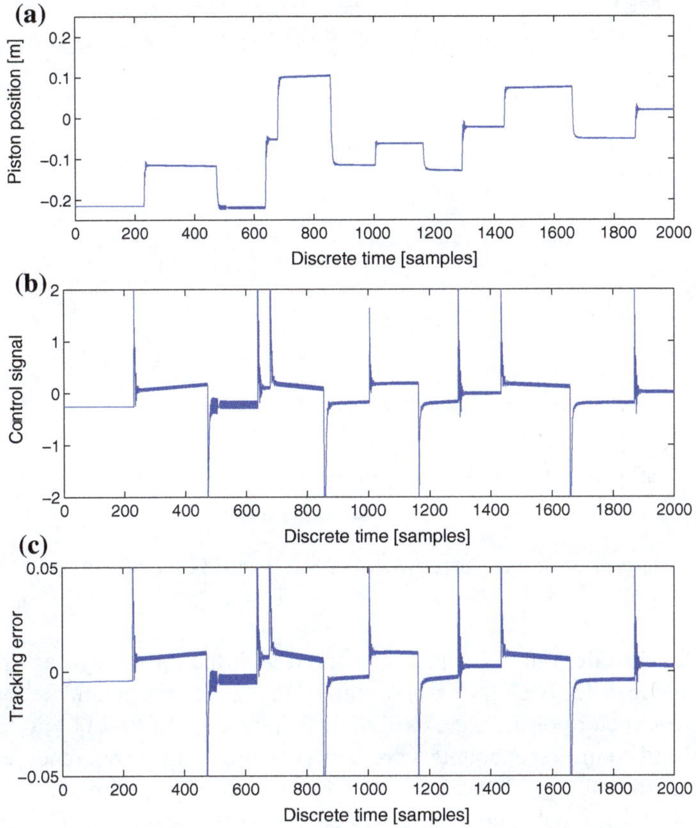

Fig. 6.12 An example interval of training data: the output signal (**a**), the control (**b**) and the tracking error (**c**)

network controller was preliminary trained using the data recorded during an evaluation of the closed-loop control with the feedback controller (see Chap. 4, Sect. 4.6). An example interval of training data is presented in Fig. 6.12. The structure of the neural controller was selected using the trial-and-error method. The best settings were set as follows: $v_c = 5$, $n_c = 2$, $n_d = 2$, $\sigma = \tanh$. The training was carried out off-line for 100 steps using the LM algorithm. The designed controller does not affect the stability of the closed-loop control because according to Theorem 6.11:

$$\|\boldsymbol{W}_1^x \boldsymbol{W}_2^x\| = 0.7965 < 1. \tag{6.96}$$

6.9.2 ILC Synthesis

The reference trajectory (in Fig. 6.13 represented by the black-solid line) was formed using the ramp signals with different slopes. Firstly, the system was investigated

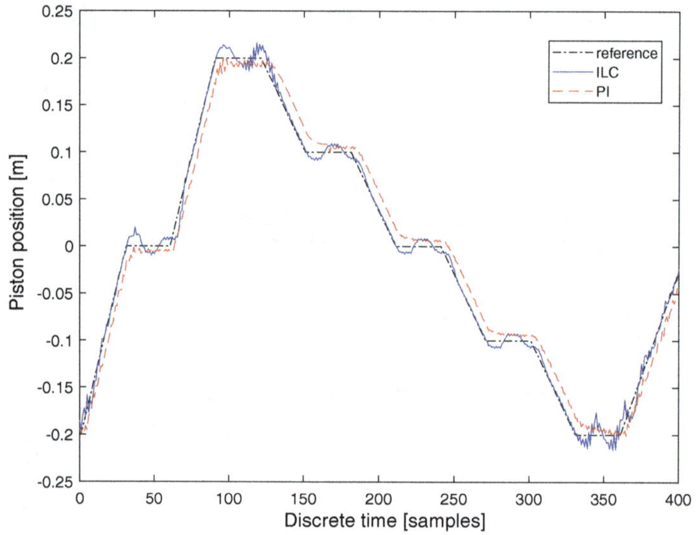

Fig. 6.13 The reference tracking: the reference (dash-dot), the PI controller (dashed), ILC (solid)

with the PI controller only. The results marked with the light-grey-dashed line are presented in Fig. 6.13. Each trial was examined using the criterion in the form of the tracking error vector norm. In the case of the PI, the control $\|e(k)\| = 0.5057$. It is clear that the PI controller encounters problems with tracking the reference, especially at the decreasing slopes of the reference. In the second experiment, the classical control system with the PI controller was enhanced by means of ILC according to the scheme shown in Fig. 6.3. The results of reference tracking (the solid line) are presented in Fig. 6.13. The results achieved by ILC are much better than those in the PI controller (dashed) case. It is important to note that after each trial, the neural network controller is retrained according to the algorithm discussed in Sect. 6.7.3. In order to avoid over-fitting, only a small number of training epochs (5 in this case) is considered. In terms of the quality of the control, it is both training parameters η and λ that had a significant impact. Through a series of experiments, the well-performing values were found to be $\eta = 0.1$ and $\lambda = 0.0001$. Figure 6.14 shows how the quality of the control (in the form of the tracking error norm) is improved over trials. At the beginning improvement is significant but later on we observe a slow convergence of the norm. This is a direct consequence of the gradient descent used as the training algorithm. Moreover, in neural network training, one can frequently observe that the training parameters selected in an optimal way at the beginning of the training process may turn out to have improper values later on. For example, an exceedingly large value of the training rate η can deteriorate the control quality for the forthcoming trials. In order to avoid such problems in cases when the value of the error norm calculated for the current trial is larger that the norm derived for the previous trial, the network weights are restored from the previous trial. Yet

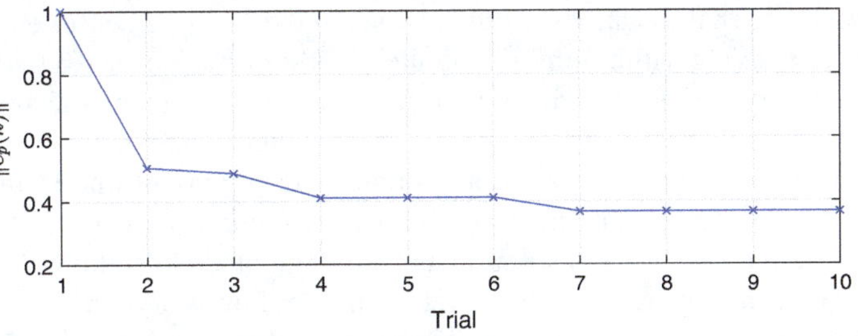

Fig. 6.14 The norm of the tracking error over 10 trials

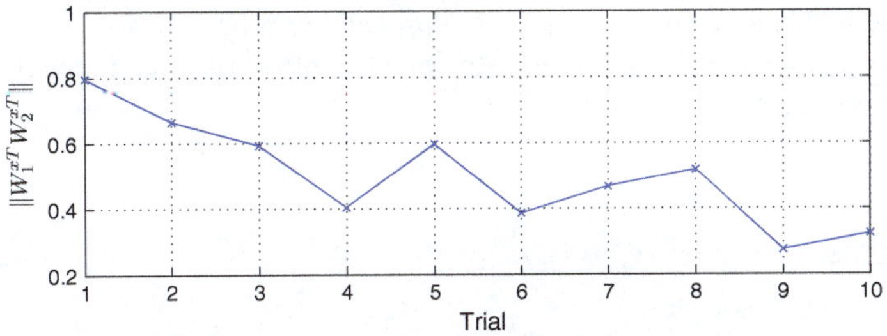

Fig. 6.15 The stability results: the values of the criterion (6.92) along trials

another important problem is to keep the neural network controller stable after each retraining. As the initial controller is a stable one, as indicated in the Sect. 6.9.1, our goal is to satisfy the criterion (6.92) after each retraining of the neural controller. This problem can be easily solved by checking the stability of the controller after each training epoch. If new network weights derived by the training procedure satisfy the condition (6.92) they are accepted, otherwise they are rejected and the controller keeps its own old weight parameters. The values of the criterion (6.92) over 8 trials are depicted in Fig. 6.15. It is obvious that the neural network controller is stable all the time.

6.10 Concluding Remarks

The chapter proposed a novel approach to the problem of designing iterative learning control with the use of neural networks. Clearly, the presented control schemes constitute an attractive alternative to the conventional adaptive control schemes for repetitive processes. An undeniable advantage of the proposed approach is the simplicity of the training algorithm and the flexibility in applying it to the case of nonlinear systems, especially with non-invertible dynamics. In this chapter, two solutions of the learning controller are proposed. The static controller is designed by means of the feed-forward neural network while the dynamic controller by means of a neural network with external dynamics. Both versions of the learning controller are trained using very simple algorithm minimizing the tracking error norm. To perform training, it is necessary to determine the output sensitivities of the plant with respect to the control signals. For these purposes, another neural network, which acts as a model of the plant, is used. Our contribution, which is an important one in this context, is the formulation of the sufficient conditions for the convergence of the iterative learning scheme to the optimal solution and a careful analysis of how to exploit this result during a training process of a neural controller. For the dynamic controller, the trial-to-trial stability is also provided. The presented experiments show that neural modelling can be successfully applied in the context of iterative learning control, leading to significant improvements of the control system performance. Clearly, the undeniable advantages of the proposed approach are related to a greater flexibility of the neural controller in adapting to the plant nonlinearities and the simplicity of the developed training algorithm.

References

1. Arimoto, S., Kawamura, S., Miyazaki, F.: Bettering operation of robots by learning. J. Robot. Syst. **1**(2), 123–140 (1984)
2. Bristow, D.A., Tharayil, M., Alleyne, A.G.: A survey of iterative learning control: a learning-based method for high-performance tracking control. IEEE Control Syst. Mag. **26**(3), 96–114 (2006)
3. Chen, Y., Wen, C.: Iterative Learning Control. Convergence, Robustness, Applications. Lecture Notes in Control and Information Sciences, vol. 248. Springer, London (1999)
4. Chi, R., Hou, Z.: A new neural network-based adaptive ILC for nonlinear discrete-time systems with dead zone scheme. J. Syst. Sci. Complex. **22**, 435–445 (2009)
5. Chow, T.W.S., Li, X.D., Fang, Y.: A real-time learning control approach for nonlinear continuous-time system using recurrent neural networks. IEEE Trans. Ind. Electron. **47**, 478–486 (2000)
6. Freeman, C., Hughes, A.M., Burridge, J., Chappell, P., Lewin, P., Rogers, E.: Iterative learning control of FES applied to the upper extremity for rehabilitation. Control Eng. Pract. **17**(3), 368–381 (2009)
7. Freeman, C.T., Rogers, E., Hughes, A., Burridge, J.H., Meadmore, K.L.: Iterative learning control in health care: electrical stimulation and robotic-assisted upper-limb stroke rehabilitation. IEEE Control Syst. **32**(1), 18–43 (2012)
8. Gao, F., Yang, Y., Shao, C.: Robust iterative learning control with applications to injection molding process. Chem. Eng. Sci. **56**(24), 7025–7034 (2001)
9. Havlicsek, H., Alleyne, A.: Nonlinear control of an electrohydraulic injection molding machine via iterative adaptive learning. IEEE/ASME Trans. Mechatron. **4**(3), 312–323 (1999)
10. Haykin, S.: Neural Networks. A Comprehensive Foundation, 2nd edn. Prentice-Hall, New Jersey (1999)
11. Kim, D.I., Kim, S.: An iterative learning control method with application for CNC machine tools. IEEE Trans. Ind. Appl. **32**(1), 66–72 (1996)
12. Kowalów, D., Patan, M.: Optimal sensor selection for model identification in iterative learning control of spatio-temporal systems. In: 21st IEEE Conference on Methods and Models in Automation and Robotics (MMAR), pp. 70–75 (2016)
13. Lee, J.H., Lee, K.S.: Iterative learning control applied to batch processes: an overview. Control Eng. Pract. **15**(10), 1306–1318 (2007)

14. Mezghani, M., Roux, G., Cabassud, M., Le Lann, M.V., Dahhou, B., Casamatta, G.: Application of iterative learning control to an exothermic semibatch chemical reactor. IEEE Trans. Control Syst. Technol. **10**(6), 822–834 (2002)
15. Moore, K.L.: Iterative Learning Control for Deterministic Systems. Advances in Industrial Control. Springer, London (1993)
16. Nelles, O.: Nonlinear System Identification. From Classical Approaches to Neural Networks and Fuzzy Models. Springer, Berlin (2001)
17. Owens, D.H., Daley, S.: Iterative learning control - monotonicity and optimization. Int. J. Appl. Math. Comput. Sci. **18**, 279–293 (2008)
18. Paszke, W., Rogers, E., Gałkowski, K., Cai, Z.: Robust finite frequency range iterative learning control design and experimental verification. Control Eng. Pract. **21**(10), 1310–1320 (2013)
19. Patan, K.: Neural network based model predictive control: fault tolerance and stability. IEEE Trans. Syst. Control Technol. **23**(3), 1147–1155 (2015)
20. Patan, K., Patan, M.: Design and convergence of iterative learning control based on neural networks. In: Proceedings of the European Control Conference, ECC 2018, Limassol, Cyprus (2018)
21. Patan, K., Patan, M., Kowalów, D.: Neural networks in design of iterative learning control for nonlinear systems. In: IFAC Papers On-line, 20th IFAC World Congress, Toulouse, France, vol. 50, pp. 13402–13407 (2017). https://doi.org/10.1016/j.ifacol.2017.08.2277
22. Quinn, S.L., Harris, T.J., Bacon, D.W.: Accounting for uncertainty in control-relevant statistics. J. Process Control **15**, 675–690 (2005)
23. Rogers, E., Galkowski, K., Owens, D.H.: Control Systems Theory and Applications for Linear Repetitive Processes, vol. 349. Springer, Berlin (2007)
24. Seel, T., Schauer, T., Raisch, J.: Monotonic convergence of iterative learning control systems with variable pass length. Int. J. Control **90**, 393–406 (2017)
25. Shen, D., Zhang, W., Xu, J.: Iterative learning control for discrete nonlinear systems with randomly iteration varying lengths. Syst. Control Lett. **96**, 81–87 (2016)
26. Tayebi, A., Abdul, S., Zaremba, M., Ye, Y.: Robust iterative learning control design: application to a robot manipulator. IEEE/ASME Trans. Mechatron. **13**(5), 608–613 (2008)
27. Uchiyama, M.: Formulation of high-speed motion pattern of a mechanical arm by trial. Trans. SICE (Soc. Instrum. Contr. Eng.) **14**(6), 706–712 (1978)
28. Xiong, W., Ho, D.W.C., Yu, X.: Saturated finite interval iterative learning for tracking of dynamic systems with hnn-structural output. IEEE Trans. Neural Netw. Learn. Syst. **27**, 1578–1584 (2016)
29. Xu, J.X., Tan, Y.: Linear and Nonlinear Iterative Learning Control for Deterministic Systems. Lecture Notes in Control and Information Sciences, vol. 291. Springer, Berlin (2003)
30. Yang, D.R., Lee, K.S., Ahn, H.J., Lee, J.H.: Experimental application of a quadratic optimal iterative learning control method for control of wafer temperature uniformity in rapid thermal processing. IEEE Trans. Semicond. Manuf. **16**(1), 36–44 (2003)

Chapter 7
Concluding Remarks and Further Research Directions

Owing to their powerful properties, artificial neural networks have become a popular choice in terms of problems occurring in control theory. Furthermore, the following two trends in modern control: robust control and fault-tolerant control can be effectively realized using appropriate neural-network architecture. This monograph is devoted to the selected designs of the robust and fault-tolerant control systems for nonlinear processes. The reported approaches mainly use the capability of a neural network to learn from historical data and to approximate nonlinear functions with an assumed accuracy. These two properties are extremely useful when dealing with nonlinear industrial plants for which a mathematical model is unknown or is very expensive to determine. Moreover, we note that even if a mathematical model exists, it is not a perfect replica of the plant dynamics. Therefore, there always exists uncertainty associated with the model, which is described by an unknown relation of the process variables. In all these cases, neural networks constitute a valuable and interesting alternative in comparison to the classical approaches.

In the light of the presented material, the original contribution reported in this monograph was to develop efficient algorithms able to design robust and fault-tolerant control schemes for nonlinear systems. In order to accomplish this task, a relevant theoretical background was also developed. Apart from new results, a number of known methods were generalized. The following is a summary of the original contributions provided by this monograph in terms of the state-of-the-art robust and fault tolerant control theory:

- Developing nonlinear model predictive control based on a neural-network predictor and constrained optimization. Formulating stability conditions of the described predictive control scheme showing that the cost is monotonically decreasing in time. Incorporating the stability conditions as the terminal constraints of the optimization problem. Analyzing fault tolerance properties of the considered NMPC. Developing sensor fault-tolerant control using multi-valued diagnostic matrices and fault estimation.
- Developing robust model predictive control using neural-network predictor and uncertainty modelling. In order to facilitate a control sequence calculation, the

© Springer Nature Switzerland AG 2019

K. Patan, *Robust and Fault-Tolerant Control*, Studies in Systems, Decision and Control 197, https://doi.org/10.1007/978-3-030-11869-3_7

so-called instantaneous linearization of a nonlinear model is applied. In such a case, the optimization problem boils down to the quadratic one. Developing the uncertainty model by means of model error modelling based on neural networks. Formulating stability conditions of the described robust predictive control scheme.

- Developing an active reconfigurable control scheme using neural network based fault accommodation. The proposed scheme enhances the existing feedback control with fault-tolerant properties. The proposed approach uses the neural model representing the system in the normal operating conditions and neural observer able to estimate the system state in case of faults. In such a setting, it is possible to estimate a fault function and calculate a correction control signal. A stability analysis of the overall control system is also carried out. Two cases are considered: the P as well PI feedback controllers.
- Developing nonlinear iterative learning control based on neural networks. The proposed approach is designed for nonlinear systems working in the repetitive regime. The idea is to propose a nonlinear learning controller, whose parameters are adapted after each trial. A convergence analysis of the proposed control scheme is carried out for the controller of the static structure. The stability along trial axis is proved for the dynamic learning controller. The learning algorithm of the controller is provided based on the gradient descent and weight decay technique.

All proposed approaches were verified on a number of case studies including simulated as well as real-world plants. The following applications were discussed:

- a tank unit simulator
  - a fault-tolerant MPC based on neural networks (Sect. 4.4),
  - fault accommodation and control reconfiguration based on a neural model and a neural observer (Sect. 5.6).
- a pneumatic servomechanism simulator
  - robust MPC based on neural networks (Sect. 4.6),
  - robust MPC with statistical bounds (Sect. 4.8),
  - nonlinear ILC with the static neural controller (Sect. 6.5),
  - nonlinear ILC with the dynamic neural controller (Sect. 6.9).
- a two-tank laboratory stand
  - a fault-accommodation and control reconfiguration based on neural networks (Sect. 5.7).
- a magnetic suspension laboratory stand
  - a nonlinear ILC with the static neural controller (Sect. 6.6).

From the engineering point of view, the proposed approaches can be regarded as a strong alternative to popular and widely used linear control schemes, the disadvantages of which becoming evident in the context of nonlinear processes. As a matter of fact, the portrayed solutions are more complex than the classical robust and fault-tolerant control schemes as they need certain calculations to be performed on-line, e.g. a retraining of neural models, the optimization of the cost function, an instantaneous linearization or stability verification by means of LMI. However, many of these calculations can be carried out using efficient numerical procedures. Moreover, micro-controllers currently in use in process-control systems have both the processing power and memory resources sufficient for a successful implementation of the control strategies proposed in the book. Therefore, we strongly believe that these advantages establish a strong argument for the use of the discussed methodologies in terms of broadly conceived engineering applications. Obviously, there is still room for refinements. A number of issues require closer attention and thus indicate further research directions. In particular, the following research problems can be pointed out:

- to implement more efficient algorithms for solving nonlinear optimization with constraints,
- to investigate an actuator fault-tolerant MPC,
- to develop a method for selecting the optimal training data for estimating the uncertainty of the nonlinear model of a plant,
- to formulate constraints imposed on the augmented control ensuring the stability of the fault-accommodation and control-reconfiguration scheme,
- to find less restrictive convergence conditions for nonlinear ILC,
- to investigate nonlinear version of a high order ILC,
- to combine MPC with ILC in order to achieve a satisfactory tracking performance in the presence of model uncertainty and disturbances.

Index

© Springer Nature Switzerland AG 2019

K. Patan, *Robust and Fault-Tolerant Control*, Studies in Systems, Decision and Control 197, https://doi.org/10.1007/978-3-030-11869-3

I

K

L

M

N

CPSIA information can be obtained
at www.ICGtesting.com
Printed in the USA
LVHW082026210319
611419LV00002B/10/P

9 783030 118686